Practical Aspects of Animal Biotechnology and Immunology

THE AUTHOR

Dr. Sunita K. Yadav is an Assistant Professor at Daulat Ram College at the University of Delhi in India. She earned her Ph.D. from the School of Biotechnology at Jawaharlal Nehru University and graduate and post-graduate degrees from the University of Delhi. She has ten years of teaching experience, during which time she has taught courses in Genetics, Immunology, and Biotechnology. She has served as the college magazine's coordinator for two years in a row and has attended numerous National and International conferences. She has published numerous research publications in high impact journals such as fish and shellfish immunology, microbial pathogenesis, and journal of medical virology. In order to motivate students for research, she has been organizing various intercollege workshops on scientific writing and research analyses tools and techniques.

Practical Aspects of Animal Biotechnology and Immunology

— Author-cum-Editor —

Dr. Sunita K. Yadav

Zoology Department, Daulat Ram College
University of Delhi, India

2024

Daya Publishing House®
A Division of
Astral International Pvt. Ltd.
New Delhi – 110 002

ISBN 978-93-5919-842-2 (HB)

Published by : **Daya Publishing House®**
A Division of
Astral International Pvt. Ltd.
– ISO 9001:2015 Certified Company –
4736/23, Ansari Road, Darya Ganj
New Delhi-110 002
Ph. 011-43549197, 23278134
E-mail: info@astralint.com
Website: www.astralint.com

Printed at : **Replika Press Pvt Ltd**

Practical Aspects of Animal Biotechnology and Immunology

— Author-cum-Editor —

Dr. Sunita K. Yadav

Zoology Department, Daulat Ram College
University of Delhi, India

2024

Daya Publishing House®
A Division of
Astral International Pvt. Ltd.
New Delhi – 110 002

ISBN 978-93-5919-842-2 (HB)

Published by : **Daya Publishing House®**
A Division of
Astral International Pvt. Ltd.
– ISO 9001:2015 Certified Company –
4736/23, Ansari Road, Darya Ganj
New Delhi-110 002
Ph. 011-43549197, 23278134
E-mail: info@astralint.com
Website: www.astralint.com

Printed at : **Replika Press Pvt Ltd**

Acknowledgements

It gives me great pleasure to thank everyone who contributed to the creation of this book in its current form. First of all, I would like to thank my mentor and guide, Prof. Aparna Dixit, whose teachings are truly invaluable. I am extremely grateful to my friend and colleague, Dr. Shailesh Kumar R. Yadav, for reviewing the book unconditionally. I would also like to thank Dr. Sonika Sharma, Dr. Madhu, Dr. Sushma Bhardwaj, and Dr. Shailesh Kumar R. Yadav for contributing their sound knowledge of biotechnology and immunology in writing few chapters of the book. I would also like to acknowledge Aakriti Singh for deciphering and turning my scribbles and ideas into visually pleasing diagrams. I am immensely grateful to Dr. Nitin Bhardwaj for providing beautiful pictures of the apparatus. I also want to thank Astral publications' editorial team for their crucial assistance. I appreciate the patience and support of my family, especially Ravi, Pooja, Milap, and Nupur, for bearing with me throughout the writing of this manuscript and the support provided in the form of fun outings to break the monotony of work and to rejuvenate me. The motivation to explore something new every day is invigorated by my little boys, Pratyaksh and Nakshatra and my niece, Anaya. I love them to the moon and back. Finally, I want to express my gratitude to my parents (including mother-in-law) for their unconditional love and support as I worked on the book for many hours. I owe so much to them.

I'd also like to thank the Principal, Dr. Savita Roy, and all of my colleagues and students at Daulat Ram College for their unwavering support and dedication. The main framework of this book is based on their curiosity, comments, and feedback over the years.

Sunita K. Yadav

Preface

The areas of Biotechnology and Immunology have always been fascinating and touch many aspects of our lives. The emergence of Biotechnology and Immunology as important areas of applied sciences coincided with a growing demand for texts that can be referred by a variety of life science courses. The students, new to these fields would require the basic and essential concepts of biotechnology and immunology. However, most of the book texts are either difficult or available in extensive detail. In my teaching career till date, I have not come across a practical book that can satisfy all the practical aspects of the Biotechnology and Immunology curriculum prescribed in undergraduate course-work of Zoology and other life sciences fields. Therefore, I have attempted to create a practical textbook that contains all the exercises of Biotechnology and Immunology prescribed in the undergraduate course-work of Zoology at the University of Delhi.

Over the past decade, there have been numerous advances in biotechnology and immunology. Despite these advances, many common techniques still remain the same. The basic essence of biotechnology revolves around DNA, restriction enzymes, and a few important biotechnological techniques. *E. coli* is considered an important model organism that is easily grown and worked upon without any ethical approval. Hence, in the various experiments in biotechnology included in the book, *E. coli* is used. In the restriction mapping chapter, many problems are incorporated with colourful restriction maps together with illustration of fragments on an agarose gel and a step-by-step guide to drawing a restriction map for easy understanding. For immunological experiments, a simplified version of all the experiments and technologies is incorporated that would be easy to comprehend. For experiments where mathematical calculations are required, a solved example is included. Since the book is mainly targeted at undergraduate students, terms are explained within the text and a comprehensive view is provided that includes abstract, introduction, principle, procedure, result, and discussion, along with precautions, and at the end of each chapter, questions are included to assess the understanding of the learners. The Animal Biotechnology and Immunology subjects are best understood when the teaching material is accompanied by colourful creations. To provide a thorough understanding of each practical, we have included many well-labelled pictures and diagrams. The book includes more than 60 colourful conceptual pictures or diagrams to make it visually appealing and student-friendly. There is a structured layout presented in the book that can be easily comprehended by the students. The detailed procedure is broken down into multiple steps along with a flow-chart that will help the learners easily understand and memorize the concept. The abstract describes the foundation of the chapter and the learning objectives will assist students understand what they will obtain from the chapter. The principles of each experiment have been explained in detail to provide the fundamentals of the technique. Many results (wherever we could) are shown in the form of diagrams or pictures that are obtained while doing the experiment for a comprehensive explanation of the observation. To introduce safe laboratory practises to undergraduate students, an introductory paragraph with points on the Dos and Don'ts in the laboratory is included. We hope that this book would be beneficial for the students to understand and expand their knowledge of biotechnology and immunology and will also help other universities and academic institutions to establish courses in biotechnology and immunology.

Sunita K. Yadav

Contents

Part B: IMMUNOLOGY

The Contributors

AUTHOR-CUM-EDITOR

Dr. Sunita K. Yadav

Assistant Professor, Department of Zoology, Daulat Ram College, University of Delhi, New Delhi-110007, India, e-mail: sunitayadav@dr.du.ac.in

OTHER CONTRIBUTORS

Dr. Sonika Sharma

Assistant Professor, Department of Zoology, Daulat Ram College, University of Delhi, New Delhi-110007 India, e-mail: sonikasharma@dr.du.ac.in

Dr. Madhu

Assistant Professor, Department of Zoology, Daulat Ram College, University of Delhi, New Delhi-110007 India, e-mail: madhu@dr.du.ac.in

Dr. Shailesh Kumar R. Yadav

Assistant Professor, Department of Zoology, Ramjas College, University of Delhi, New Delhi-110007, India, e-mail: shailesh2346@gmail.com

Dr. Sushma Bhardwaj

University of Delhi, New Delhi-110007, India, e-mail: sushma.bhardwaj@gmal.com

Aakriti Singh

Department of Botany, University of Delhi, New Delhi-110007, India, e-mail: aakritisingh17nov@gmail.com

Dr. Nitin Bhardwaj

Assistant Professor, Immunology Laboratory, Gurukul Kangri University, Haridwar, Uttarakhand, e-mail: nitin.bhardwaj@gkv.ac.in

Abbreviations

%	Percentage
×	Times
µg	Microgram(s)
µl	Microlitre(s)
µM	Micromolar
A	Absorbance
Ab	Antibody
Ag	Antigen
ALS	Alkaline lysis solution (I,II,III)
APC	Antigen processing cell
APS	Ammonium persulphate
ATP	Adenosine triphosphate
BALT	Bronchus associated lymphatic tissues
BCIP	5-bromo-4-chloro-3'-indolyphosphate p-toluidine
bp	Base pairs
BSA	Bovine serum albumin
CFU	Colony forming units
CI	Chloroform Isoamylalcohol
Da	Dalton
DD	Double distilled
DNA	Deoxyribonucleic acid
dNTPs	Deoxyribonucleotide triphosphates
EDTA	Ethylenediaminetetraacetic acid
ELISA	Enzyme-linked immunosorbent assay

EtBr	Ethidium bromide
F	Forward
g	Gram(s)
GALT	Gut associated lymphocyte tissue
h	Hour(s)
HRP	Horseradish peroxidase
IgG	Immunoglobulin G
IgM	Immunoglobulin M
kb	Kilobase pairs
kDa	Kilodalton(s)
L	Litre
LB	Luria-Bertani
LPS	Lipopolysaccharide
M	Molar
MALT	Mucous associated lymphoid tissues
mg	Milligram(s)
$MgCl_2$	Magnesium chloride
$MgSO_4$	Magnesium sulfate
min	Minute(s)
ml	Millilitre(s)
mM	Millimolar(s)
N	Normal
NALT	Nasal associated lymphoid tissues
ng	Nanogram(s)
O.D.	Optical density
O/N	Overnight
ºC	Degree Celsius
PAGE	Polyacrylamide gel electrophoresis
PBS	Phosphate buffered saline
PBST	Phosphate buffer saline tween-20
PCR	Polymerase chain reaction
pH	$-\log[H^+]$
pI	Isoelectric point

PVDF	Polyvinylidenedifluoride
R	Reverse
RE	Restriction enzyme
RES	Restriction enzyme site
RFLP	Restriction fragment length polymorphism
RNA	Ribonucleic acid
rpm	Revolution per minute
RT	Room temperature
SDS	Sodium dodecyl sulfate
sec	Seconds
SSR	Simple sequence repeats
TBE	Tris-boric acid
TE	Tris-EDTA
TEMED	Tetramethylethylenediamine
Tm	Midpoint of temperature denaturation curve
U	Unit
UV	Ultraviolet light
V	Volt(s)
v/v	Volume/volume
VNTR	Variable number tandem repeats
W	Watt
w/v	Weight/volume
w/w	Weight/weight
α	Alpha
β	Beta
λ	Lamda

General Operating Procedures

1. Enzymes: Enzymes are often costly and can be damaged by extended room temperature exposure or contamination. For each enzyme, use a new, sterile pipette tip. Never keep the enzyme at room temperature for long. The enzymes should always be added last in a reaction; also adding enzymes to unbuffered solutions can inhibit their activity.
2. Pipette tips: Sterile tips should be used while working with DNA, proteins or enzymes. The tip box should be closed immediately after using the tips. In case, if tips get contaminated, they should be autoclaved and reuse.
3. Equipment: The usage and general precautions about each equipment should be known before attempting to operate the equipment. The laboratory equipment are generally very expensive and in case of damage, the cost can be exorbitantly high, also inappropriate usage of the equipment can be extremely dangerous.
4. Lab Safety: All the general lab safety guidelines should be followed.

Safety in Laboratories

The exposure to hazardous substances and ultraviolet (UV) irradiation should be minimal while working in the laboratory. To minimize the exposure to these hazards along with recombinant DNA, the following regulations will be carefully enforced to guarantee the safety and well-being of students and support workers. These rules should be followed by everyone and at all times in the laboratory:

- ☆ No eating or drinking is permitted in the laboratory. Food and beverages should not be brought into the lab.
- ☆ In the laboratory, safety glasses must be worn at all times and put on a lab coat and long pants.
- ☆ No open-toed shoes or sandals are permitted in the laboratory at any time.
- ☆ When working with any hazardous or possibly dangerous substance, always wear gloves. To avoid contamination of equipment, *etc.*, change gloves periodically and remove them before leaving the laboratory.
- ☆ To prevent the dangers of open flames, cultures, and dangerous chemicals, long hair must be tied and kept back at all times and loose-fitting clothing avoided.
- ☆ The microorganisms when not used, should be placed in bags together with their tubes required for growth. The bags should be properly labelled and autoclaved. Bleach can be added to the microorganism-containing liquid medium.
- ☆ Remove all superfluous items from your lab bench. Personal things and extra supplies can be stored in cabinets and drawers. Clean the working area and switch off the equipment when work is over for the day.

- ☆ When using the transilluminator without safety cabinet, a UV-protective full-face shield should be worn. Do not expose yourself in front of transilluminator for long to analyse the gel, instead click a picture of the gel to study it.
- ☆ Make it a habit to wash hands thoroughly before leaving the laboratory.
- ☆ Try to not spill anything by working carefully. If still there is any spillage or accident (injury), it should be cleaned up as soon as possible or notify the laboratory staff.
- ☆ Everyone should be aware of the location of first-aid box, fire extinguisher, wash basin, and emergency exits.

General Helpful Tips while Working in the Lab

- ☆ Sterile water should be used for all the experiments to prevent contamination. Tap water should not be used.
- ☆ Pipette tips should be changed between each sample/reagent.
- ☆ Mix the samples well but gently during each experiment
- ☆ Wipe the work area with 70 per cent alcohol before starting the experiment.

Part A

Animal Biotechnology

1

Isolation of Genomic DNA from *E. coli*

Dr. Sunita K. Yadav

ABSTRACT

Genomic DNA is the heritable genetic information stored in each cell that gives the information needed to develop, maintain and reproduce an organism. In eukaryotic cells, genomic DNA is located in the nucleus, organised in the form of chromosomes, protected by proteins called histones. The capacity to extract and purify genomic DNA from a variety of biological samples is a crucial initial step in the investigation of many biological processes. Purification of genomic DNA requires lysis or homogenization of cell/tissue to release genomic DNA from nucleus followed by separation of cell debris and other cellular material. All the steps, along with the precipitation of genomic DNA, are carried out with the help of different chemicals. The current chapter is focused on the purification of genomic DNA from E. coli. The principle involving role of various chemicals is explained along with flow chart for easy understanding.

Keywords: *LB medium, GTE mix, agarose gel, DNA fragment, E. coli*

Learning Objectives

After the experiment, students will be able to:

- ☆ Isolate the genomic DNA from a given sample
- ☆ Understand the role of various chemicals in genomic DNA isolation
- ☆ Explain the importance of genomic DNA isolation in various applications

1. Introduction

About *E. coli*

E.coli is a gram negative bacterium with a cell wall composed of an outer thick layer of lipopolysaccharides and inner thin layer of peptidoglycans. The size of the *E. coli* genome is 4.6 MB and resides in the form of double stranded closed circular DNA. It generally has one or more extrachromosomal DNA in the form of plasmids. *E. coli* is considered a model organism and is usually employed in the laboratory for various experiments.

2. Principle behind Genomic DNA Isolation

Genomic DNA isolation is generally the first step required for cloning and in various other applications. The isolation of genomic DNA from bacteria is a simple process based on the differential nature of the molecules present in bacteria. The *E. coli* should be grown at optimum temperature (37°C) and medium (Luria-Bertani broth) conditions and needs to be harvested when the O.D. at 600 nm reaches 1-1.2 for maximum yield of DNA (>0.8, late log to early stationary phase). The genomic DNA should be separated from other macromolecules of a cell such as RNA, protein, lipids, *etc.* The cell membrane of cell should be broken to release DNA along with other molecules of cell in the buffer. GTE buffer is used in the first step followed by SDS. GTE consists of glucose, Tris.HCl and EDTA. Glucose maintains the osmolarity of the DNA and Tris.HCl provides alkaline pH to preserve DNA. EDTA prevents the activity of nucleases (that can chew the DNA) by chelating essential ions such as Mg^{+2} that are required for their activity. Also removal of these ions prevents maintenance of conformation of cell wall which will be eventually lysed in the next steps. SDS is used to disrupt the cell membrane and cause denaturation of protein resulting in deproteinization of DNA. Phenol and chloroform solubilize lipids and other organic components of cell leaving DNA and RNA molecules in the aqueous phase. Also, phenol/chloroform treatment can precipitate the proteins by denaturing them. The released DNA is precipitated by using chilled ethanol or isopropanol which exposes negatively charge group of DNA. Sodium chloride binds to phosphate groups of DNA and reduces repulsion among the polynucleotide chain of DNA resulting in precipitation and hence separation of DNA.

3. Requirements

- ☆ Glasswares/plasticwares: Microfuge tubes (1.5 ml) and microtips, flasks and tubes (for culturing).
- ☆ Equipment: Agarose gel electrophoresis apparatus (Figure 1a), centrifuge, pH meter, orbital shaker, micropipettes, vortex, microwave, autoclave, UV-transilluminator.
- ☆ Biological Material: Culture of *E. coli* (DH5α)
- ☆ Media: Luria Bertani medium (LB)

☆ Chemicals/Reagents:

- ❖ For genomic DNA isolation: GTE mixture (Glucose, Tris.HCl, EDTA), SDS, NaCl, Isopropanol, 70 per cent ethanol.
- ❖ For screening on agarose gel: Agarose, TBE buffer, EtBr, Bromophenol blue dye

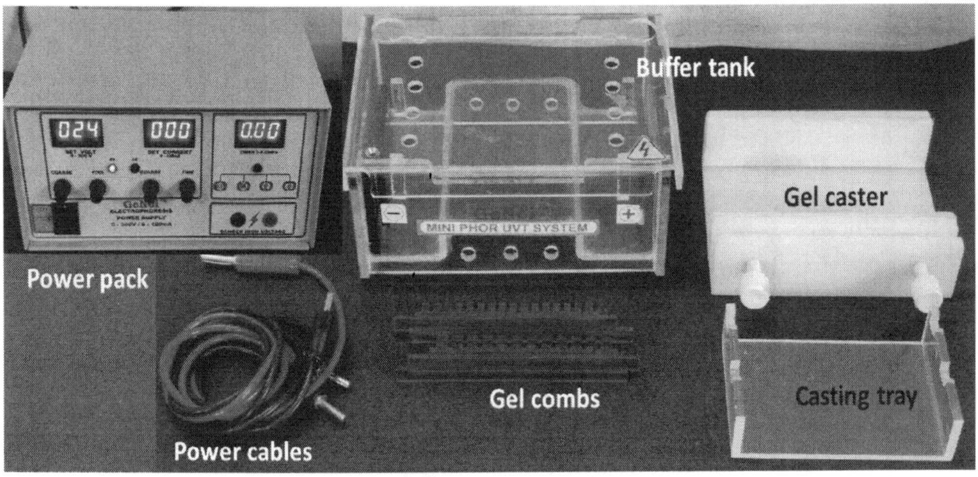

Figure 1a. Electrophoresis Equipment Showing all the Parts Required to Run the Gel (Pic courtesy: Dr. Nitin Bhardwaj).

Preparation of Chemicals

LB (Luria-Bertani) Broth, pH 7.0-7.5

Sl.No.	*Chemical*	
1.	Tryptone	10 g
2.	Yeast extract	5 g
3.	NaCl	10 g
4.	Glucose (optional)	1 g

Note: After mixing all the constituents and adjusting the pH and make up the volume to 1000 ml with distilled water, the LB containing flask should be autoclaved and can be stored at room temperature (RT).

Other Chemicals

Sl.No.	*Chemical*	*M.Wt. (g)*	*Stock*	*10 ml*	*Autoclave*	*Storage*
1.	NaOH	40.00	10 N	4 g	No	RT
2.	Glucose	180.16	1 M	1.8 g	Yes	4°C
3.	Tris.Cl	121.14	1 M	1.21 g	Yes	RT

Sl.No.	Chemical	M.Wt. (g)	Stock	10 ml	Autoclave	Storage
4.	EDTA	372.24	0.5 M	1.86 g	Yes	RT
5.	NaCl	58.44	5 M	2.92 g	Yes	4°C
6.	SDS	288.38	10 per cent (volume by volume)	10 g	No	RT
7.	EtBr	394.294	10 mg/ml	—	No	RT and dark

GTE (Glucose, Tris.HCl, EDTA, pH 8.0) buffer preparation:

Sl.No.	Chemical	Concentration	For 100 ml
1.	Glucose	50 mM	5 ml from 1 M stock
2.	Tris.HCl, pH 8.0	50 mM	5 ml from 1 M stock
3.	EDTA, pH 8.0	10 mM	2 ml from 0.5 M stock

6× Gel Loading Dye (store at -20°C)

Sl.No.	Chemical	Composition	For 10 ml	Autoclave
1.	Glycerol	50 per cent	5 ml of 100 per cent stock	No
2.	EDTA pH 8.0	0.2 M	4 ml of 0.5 M stock	No
3.	Bromophenol blue	0.25 per cent	A pinch	No

10×TE Buffer pH 8.0 (store at RT)

Sl.No.	Chemical	Composition	For 10 ml	Autoclave
1.	Tris.HCl, pH 8.0	100 mM	1 ml of 1 M stock	Yes
2.	EDTA, pH 8.0	10 mM	0.2 ml of 0.5 M stock	Yes

10× TBE Buffer pH 8.3 (store at RT)

Sl.No.	Chemical	Composition	M.Wt. (g)	For 100 ml	Autoclave
1.	Tris.HCl	900 mM	121.14	10.9 g	No
2.	Boric acid	900 mM	61.83	5.56 g	No
3.	EDTA	20 mM	372.24	0.74 g	No

4. Procedure

4.1. Isolation of Genomic DNA

1. Harvest 1.5 ml overnight (O/N) grown culture of *E. coli* by centrifugation at 5000 rpm for 5 min.
2. Add 200 µl of GTE buffer to the cell pellet and resuspend the cells in it by mixing or vortexing.
3. Add 400 µl of 1 per cent SDS to the above tube and gently mix the solution (cells will lyse at this step). Proteinase k (5 µl from 1 U/1 µl) addition is optional. Keep the tube at RT for 5 min.
4. Add 120 µl of 5 M NaCl to the tube and incubate at RT for 5 min.
5. Equal volume of phenol-chloroform (~700 µl) should be added and mix well by inverting. Incubate the tube at RT for 5 min.

   Note: This step is optional and can be skipped.
6. The contents are centrifuged at 10,000 rpm for 15 min and the aqueous phase (or supernatant in case '5' step is skipped) is separated from the organic phase by micropipette. Release the aqueous phase in to a fresh tube (proteins are separated from DNA at this step).
7. In the aqueous phase, add equal volume (~700 µl) of isopropanol. The tube is centrifuged at 10000 rpm for 15 min followed by 70 per cent ethanol wash (10000 rpm for 10 min) to remove the salts.
8. Discard all the supernatants and keep the tube at RT to dry the pellet. After the pellet is dried, add 20 µl of 1xTE buffer or double distilled water. The sample is stored for further experiments.

4.2. Agarose Gel Preparation

1. Prepare 0.8 per cent agarose solution (0.8 gm in 100 ml of 1× TBE buffer).

   (Note: The volume of agarose solution required depends upon the size of gel casting tray)
2. Heat the agarose mix in microwave oven, or in boiling waterbath, or on hot plate with continuous stirring until all particles of agarose dissolve.
3. Take a clean gel casting tray and pour the melted agarose (40°C-50°C) in it. Place the comb in the formed grooves.

   (*Note*: Casting tray should be kept horizontally without any disturbance till gel solidifies. Also,while placing comb, air bubbles should not be formed. If air bubbles are formed, remove them carefully before gel solidifies)
4. Allow the time to gel to solidify (approx. 30 min at RT).
5. Remove the comb gently without disturbing the gel.

4.3. Agarose Gel Electrophoresis (Figure 1b)

1. Place the gel casting tray in the buffer tank. Complete all the circuit attachments (anode and cathode leads attachment to respective socket of power pack).
2. Pour 1× TBE buffer into the buffer tank so that the gel should be submerged in it.
3. Add 6× gel-loading dye to the DNA samples in the final concentration of 1×.

   (*Note*: Mix 5 µl of sample + 1 µl of gel-loading dye (from 6×) on clean cello tape or directly in microfuge tube)
4. Load slowly the above dye added DNA samples to the wells of gel using micropipette.

   (*Note*: DNA marker/ladder can also be loaded to estimate the size of sample DNA. The prefered wells to load the DNA ladder are first or last)
5. Run the gel at 80-100 volt till loading dye migrates 3 quarters of the gel.

   (*Note*: DNA loaded wells should be near the cathode as DNA being negatively charged runs towards the anode)

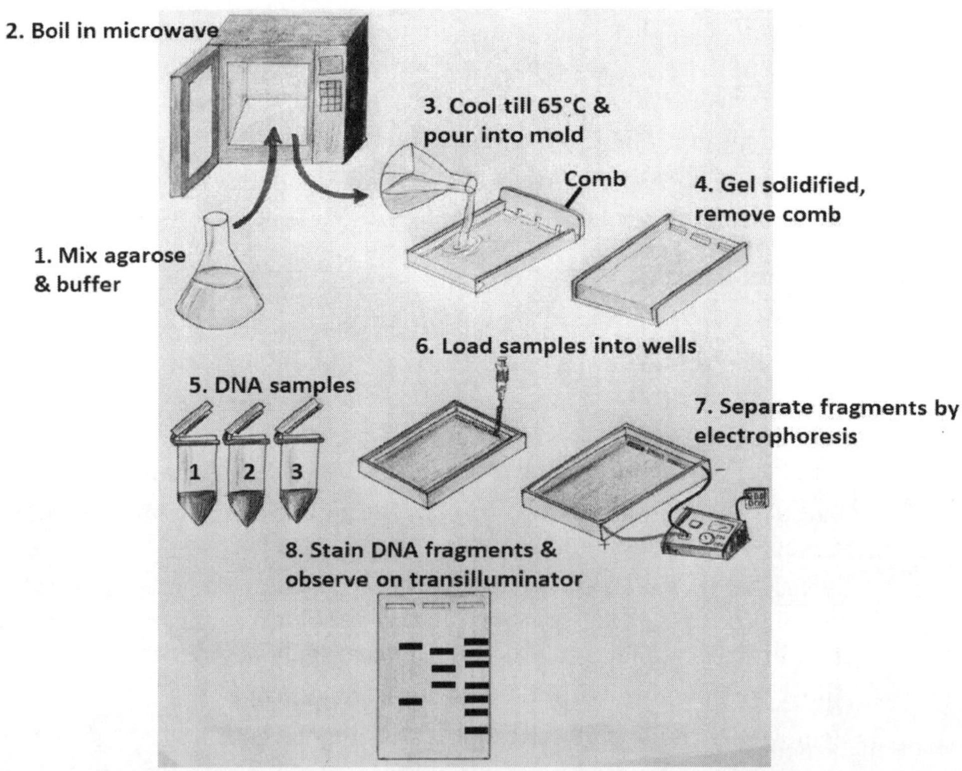

Figure 1b. Process of Agarose Gel Electrophoresis (Illustrator: Aakriti Singh, Department of Botany, University of Delhi).

6. Switch off the apparatus when DNA is sufficiently migrated.
7. Stain the gel with EtBr (0.5 μg/ml) in electrophoresis buffer for 30 min.

   (*Note*: EtBr can be directly added to the molten gel before casting it or can be added to the electrophoresis buffer while electrophoresis. Be careful with EtBr as it is carcinogenic and gloves should be worn while handling it)
8. Place the gel under a UV transilluminator and observe the DNA bands.

   (*Note*: UV rays are harmful to eyes. The protective cover/lid of the transilluminator should be closed or protective glasses should be worn while observing the gel)

Determination of the DNA Yield can be done by

UV spectrophotometry: To lower the concentration of DNA for quantification, dilute it in 1× TE buffer in 1:100 or 1:50 ratio (depending on the copy number of plasmid) and measure the absorbance (optical density) at 260 nm and 280 nm. Use 1× TE buffer for blank setting. The formula to calculate DNA concentration is:

$$\text{DNA } (\mu g/ml) = A_{260} \times \text{Dilution factor} \times 50$$

Here, 50 is the extinction coefficient of DNA.

The ratio of A_{260}/A_{280} estimates the purity of the DNA. DNA is pure if the ratio comes to 1.8.

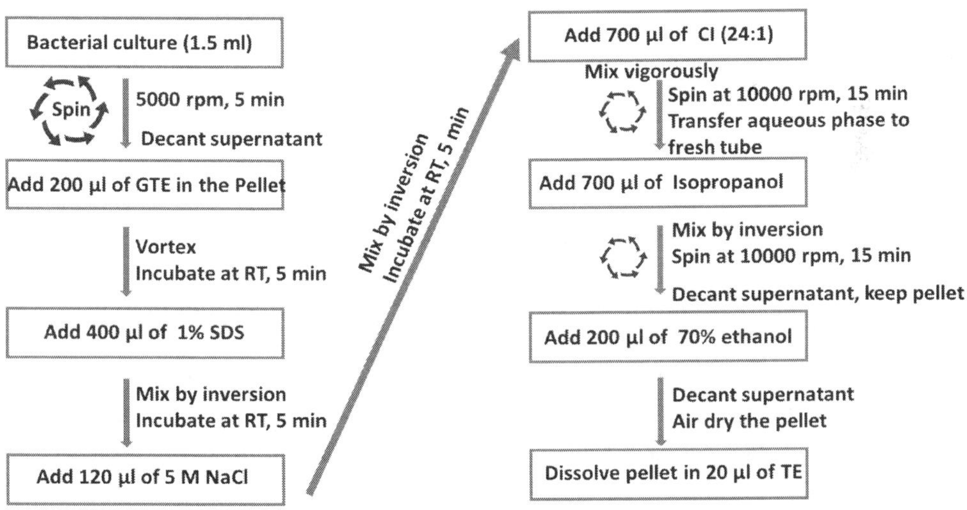

Flowchart Showing Steps of Genomic DNA Isolation.

5. Result

A high molecular weight DNA band is seen near the well. If a lambda DNA marker is also uploaded then the genomic DNA runs above 23 kb in size (Figure 1c).

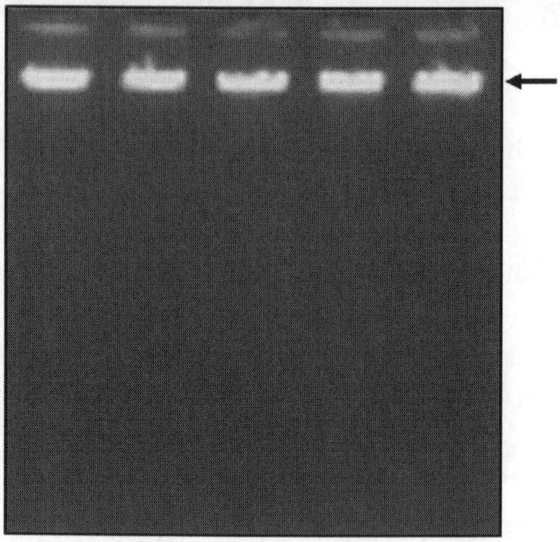

Figure 1c. All the Lanes showing Genomic DNA (marked by Arrow) Isolated from *E.coli*.

6. Discussion

The genomic DNA of bacteria is more than 1 MB in size, therefore it runs at a very slow speed on agarose gel from cathode (-) towards anode (+). During DNA isolation, it undergoes shearing but despite it, the size remains high and clearly, it can be distinguished from the low molecular size plasmids. The shearing of DNA may lead to fragmentation and a smear might be visible inplace of distinct DNA band. The plasmids and RNA can be seen on agarose gel along with genomic DNA but these are small size molecules and hence migrate at a considerably faster speed than the genomic DNA. The RNA that can interfere with the screening of genomic DNA can be removed by adding RNase which chews the RNA and is removed during the initial steps of genomic DNA isolation.

7. Precautions

- ☆ All the plastic and glassware that are being used should be clean and contamination free.
- ☆ The aqueous phase should be carefully separated from the organic phase.
- ☆ The tube content should be gently mixed after addition of SDS (step 5 of **Isolation of genomic DNA)**.
- ☆ The pellet of DNA should be air dried thoroughly before addition of TE buffer.
- ☆ The O.D. of culture should be in log phase for maximum yield of DNA.

8. Questions

Multiple Choice Questions

1. **In DNA extraction, ___is included for chelating the __ions needed for enzymes __, to prevent degradation of DNA.**
   a. EDTA, Ca^{+2}, deoxyribonuclease
   b. Enzyme, Mg^{+2}, deoxyribonuclease
   c. Chelating agent, Mg^{+2}, ribonuclease
   d. EDTA, Mg^{+2}, deoxyribonuclease
2. **The optimum O. D. for harvesting *E. coli* for genomic DNA extraction is:**
   a. < 0.4
   b. < 0.6
   c. < 0.2
   d. > 0.8
3. **The DNA is precipitated at the last step of genomic DNA isolation using:**
   a. Ethanol
   b. Isopropanol
   c. None of them
   d. Both of them
4. **In *E. coli*, genomic DNA resides as:**
   a. Single stranded closed circular DNA
   b. Double stranded closed circular DNA
   c. Single stranded open circular DNA
   d. Double stranded open circular DNA
5. **The phenol chloroform in the DNA isolation is used**
   a. To separate the RNA
   b. To separate the DNA
   c. To separate the proteins
   d. To separate the membrane

Short Answer Questions

1. **What is the principle of genomic DNA isolation?**
2. **Why do bacteria require LB medium for growth?**

Long Answer Questions

1. **What are the various constituents of LB medium? Give the function of each of them.**
2. **How do you differentiate genomic DNA and plasmid DNA on agarose gel?**

REFERENCES

Maniatis T, Fritsch EF, Sambrook J (1982) Molecular cloning: A laboratory manual. Cold Spring Harbor Laboratory, Cold Spring Harbor, NY

Becker JM, Caldwell GA, Zachgo EA (2007) Biotechnology- A laboratory course. Academic Press. An imprint of Elsevier, San Diego, California.

https: //webstor.srmist.edu.in/web_assets/srm_mainsite/files/files/BT0210 per cent 20- per cent 20MOLECULAR per cent 20BIOLOGY per cent 20LABORATORYMANUAL.pdf

Kalbande, Bipinchandra. (2012). Principle of DNA Extraction and the Role of Various Components Used in DNA Extraction. Agrobios Newsletter. 11. 6-8.

2

Isolation of Plasmid DNA (pUC 18/19) from *E. coli*

Dr. Sunita K. Yadav

ABSTRACT

Isolation of plasmid from bacteria is an important step in molecular biology techniques such as cloning and sequencing of gene, transformation, and gene therapy. Alkaline lysis is a well-established molecular biology technique that involves breaking open the cells to isolate plasmid DNA or other cell components like proteins. In this process, bacteria harbouring the desired plasmid are cultured first, then lysed in an alkaline lysis buffer solution containing a detergent, sodium dodecyl sulphate (SDS) and a strong base, sodium hydroxide. The detergent cleaves the membrane's phospholipid bilayer, while the alkali denatures the proteins that keep the cell membrane intact. The plasmid is extracted and purified by removing the cellular debris through a series of procedures that include agitation, precipitation, centrifugation, and the removal of supernatant.

Keywords: *Extrachromosomal DNA, Alkaline lysis, Pellet, Renature*

Learning Objectives

After the experiment, students will be able to:

- ☆ Isolate plasmid DNA using alkaline lysis method

- ✰ Understand the role of various chemicals in plasmid DNA isolation
- ✰ Analyse the various forms of plasmid DNA

1. Introduction

Plasmid is an extrachromosomal DNA present most commonly in bacteria and some species of Archaea and Eukarya. Plasmids have the property to replicate independently and are generally double stranded circular DNA. The term 'plasmid' was introduced by American molecular biologist, Joshua Lederberg in 1552. Although, plasmids are non-essential for bacterial growth, they contribute to the genetic diversity and evolution of bacteria. Plasmids contain antibiotic resistance genes, bio-remediation genes and genes for some protein expression. The genes present in plasmids provide advantage to plasmid harbouring bacteria over non-plasmid bacteria in certain environmental conditions. The presence of antibiotics in the environment causes death of antibiotic sensitive bacteria while antibiotic resistance gene coded by plasmid provide defense to the antibiotic resistant plasmid harbouring bacteria.

Characteristics of the Plasmid

The physical qualities of plasmids found in the bacteria vary, including size (kbp), shape, and copy number.

Plasmid Size

Plasmids can be as small as 1 kilo base pair (1000 bp) or as large as 1000 kbp megaplasmids with hundreds or thousands of base pairs.

Plasmid Shape

Despite the fact that most plasmids have a circular geometry, there are currently many occurrences of linear plasmids in bacteria. Plasmid DNA can take one of five forms: nicked open circular DNA with one strand cut, relaxed circular DNA with both strands uncut but enzymatically relaxed, linear DNA with free ends, supercoiled DNA with both strands uncut, and supercoiled denatured DNA with unpaired strands creating it less compact.

Plasmid Copy Number

The plasmid copy number denotes to the expected or average number of plasmid copies per host cell. Plasmids are classified as having a low, medium, or high copy number. The category of plasmid copy number is critical for molecular biology experiments. Though a low copy number plasmid provides a lower yield and hence many cultures are required to be grown but incase of toxic gene, the low copy plasmid prevents the lysis of bacteria in the culture. High copy number plasmids give a high yield but are not useful sometimes when the gene is toxic. The low copy number plasmids divide equally into the daughter cells during cell division in bacteria while high copy number plasmids distribute randomly.

Applications of the Plasmid

- ☆ Plasmids can be manipulated and modified to express a gene of interest (*e.g.* production of human growth hormone using genetic engineering)
- ☆ Plasmid can be used to study molecular aspects of DNA replication, conjugation and evolution.
- ☆ Plasmids are used for the production of gene libraries.

The isolation of desired plasmid is generally carried out using alkaline lysis method. The isolation process depends on the different characteristics of plasmid DNA and genomic DNA. Various cellular components of the cell such as cell wall, membrane, proteins and other cell debris are removed in sequential steps with the help of chemicals.

2. Principle

The separation of plasmid DNA from the genomic DNA of bacteria is based on the differential denaturation of them using alkaline lysis process. To isolate plasmid, cells are lysed and plasmid is separated from cellular fractions (cell debris and chromosomal DNA). The cells are resuspended first in solution I containing glucose, Tris.HCl, and EDTA. Glucose maintains the osmolarity of cells (not of DNA as it will be lysed eventually). Since DNA is acid labile, Tris.HCl provides slightly alkaline pH (8.0). EDTA inhibits DNase (can chew plasmid DNA) that is released when cell is lysed. Also, EDTA chelates magnesium ions that are no more available to maintain cell wall conformation intact. In the second step, bacterial cells are lysed with alkaline solution of NaOH containing SDS resulting in formation of cell lysate. SDS, an anionic detergent in the presence of high pH provided by NaOH, solubilizes cell membrane and denatures chromosomal DNA and proteins resulting in release of low molecular weight plasmid DNA in the supernatant. At this step, chromosomal and plasmid DNA, both are denatured; however, the closely interwined strands of plasmid DNA could not be separated from each other and hence potentially capable to renature. Treatment with potassium acetate (pH 5.0) neutralizes the alkaline medium and allows only covalent closed plasmid DNA to renature (reanneal) while chromosomal DNA (large and more complex in structure) remains denatured and aggregate in to a tangled mass. Hence, the plasmid solubilized in the solution while denatured chromosomal DNA, dismantled cell wall and proteins are coated with SDS and precipitated. Finally, all the cell debris is removed by centrifugation (Figure 2a).

3. Requirements

- ☆ Glasswares/plasticwares: Microfuge tubes (1.5 ml) and microtips, Flasks and tubes (for culturing).
- ☆ Equipment: Gel electrophoresis apparatus, centrifuge, pH meter, orbital shaker, micropipettes, vortex, Microwave, autoclave, UV-transilluminator.

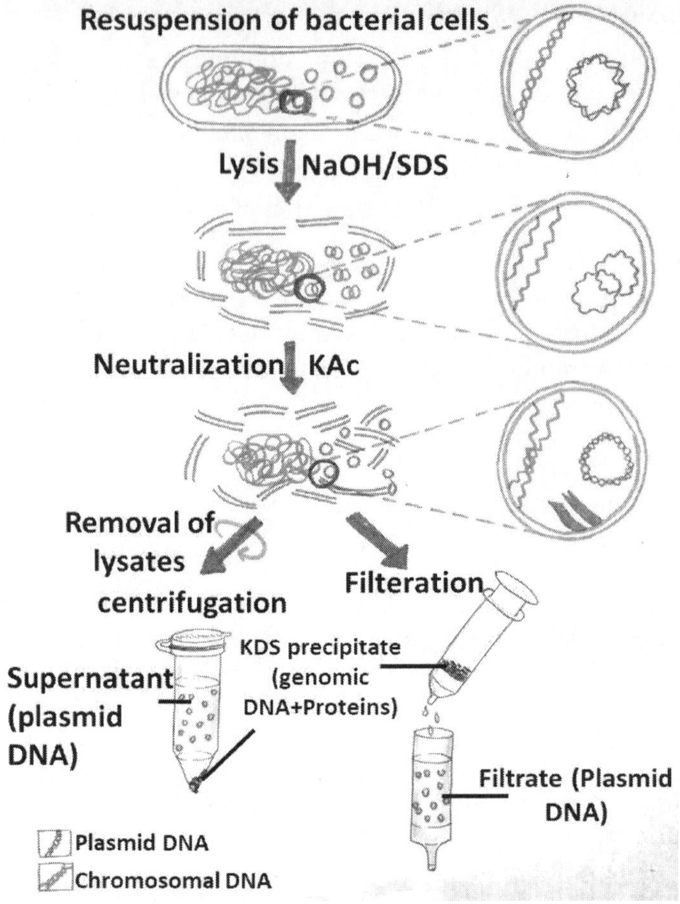

Figure 2a. Principle of Plasmid DNA Isolation
(Illustrator: Aakriti Singh, Department of Botany, University of Delhi).

- ☆ Biological Material: Culture of *E. coli* (DH5α) harbouring pUC 18 plasmid.
- ☆ *Chemicals/Reagents*

  O/N grown colony on LB plate, Microcentrifuge tubes (1.5 ml), LB broth (100 ml), Solution I (50 mM Glucose, 25 mM Tris.HCl pH 8.0, 10 mM EDTA), Solution II (1 per cent SDS, 0.2 N NaOH) (pH 12.0), Solution III (3M Sodium acetate pH 5.0), RNase (1 mg/ml), Phenol: Chloroform: Isoamyl alcohol (25:24:1), Chloroform: Isoamyl alcohol (24:1, v/v), Isopropanol, 70 per cent alcohol, agarose gel.

Buffers and Solutions

- ☆ Alkaline lysis solution I (ALSI)
- ☆ Alkaline lysis solution II (ALSII)

- Alkaline lysis solution III (ALSIII)
- Antibiotic for plasmid selection.
- Ethanol.
- TE (pH 8.0) containing 20 μg/ml RNase A.

Media

- LB medium.

Preparation of Chemicals

LB (Luria Bertini) Broth, pH 7.0-7.5

Sl.No.	Chemical	
1.	Tryptone	10 g
2.	Yeast extract	5 g
3.	NaCl	10 g
4.	Glucose (optional)	1 g

Note: After mixing all the constituents and adjusting the pH, make up the volume to 1000 ml with distilled water, the LB containing flask should be autoclaved and can be stored at RT.

Other Chemicals

Sl.No.	Chemical	M.Wt. (g)	Stock	For 10 ml	Autoclave	Storage
1.	NaOH	40.00	10 N	4 g	No	RT
2.	Ampicillin	-	100 mg/ ml	1 g	No	- 20°C
3.	Glucose	180.16	1 M	1.8 g	Yes	4°C
4.	Tris.Cl (pH 8.0)	121.14	1 M	1.21 g	Yes	RT
5.	EDTA (pH 8.0)	372.24	0.5 M	1.86 g	Yes	RT
6.	Potassium acetate	98.15	5 M	4.9 g	No	RT
7.	NaCl	58.44	5 M	2.92 g	Yes	4°C
8.	SDS	288.38	10 per cent (volume by volume)	1 g	No	RT
9.	EtBr	394.294	10 mg/ml	—	No	RT and dark

Alkaline Lysis Solution I

Sl.No.	*Chemical*	*Concentration*	*For 100 ml*
1.	Glucose	50 mM	5 ml from 1 M stock
2.	Tris.HCl, pH 8.0	25 mM	2.5 ml from 1 M stock
3.	EDTA pH 8.0	10 mM	2 ml from 0.5 M stock

Alkaline Lysis Solution II

Sl.No.	*Chemical*	*Concentration*	*For 100 ml*
1.	NaOH	0.2 N	2 ml from 1 N stock
2.	SDS	1 per cent	1 ml from 10 per cent stock

Alkaline Lysis Solution III

Sl.No.	*Chemical*	*For 100 ml*	*Autoclave*	*Storage*
1.	5 M potassium acetate	60 ml	No	RT
2.	Glacial acetic acid	11.5 ml	No	RT
3.	EDTA, pH 8.0	2 ml from 0.5 M stock	No	RT

6× Gel Loading Sye (store at -20°C)

Sl.No.	*Chemical*	*Composition*	*For 10 ml*	*Autoclave*
1.	Glycerol	50 per cent	5 ml of 100 per cent stock	No
2.	EDTA, pH 8.0	0.2 M	4 ml of 0.5 M stock	No
3.	Bromophenol blue	0.25 per cent	A pinch	No

10×TE Buffer pH 8.0 (store at RT)

Sl.No.	*Chemical*	*Composition*	*10 ml*	*Autoclave*
1.	Tris.HCl pH 8.0	100 mM	1 ml of 1 M stock	Yes
2.	EDTA, pH 8.0	10 mM	0.2 ml of 0.5 M stock	Yes

10× TBE Buffer pH 8.3 (store at RT)

Sl.No.	*Chemical*	*Composition*	*M.Wt. (g)*	*100 ml*	*Autoclave*
1.	Tris.HCl	900 mM	121.14	10.9 g	No
2.	Boric acid	900 mM	61.83	5.56 g	No
3.	EDTA	20 mM	372.24	0.74 g	No

4. Procedure

Plasmid DNA is Isolated using Alkaline Lysis Method

1. Grow bacterial culture O/N in LB containing suitable antibiotic at 37 °C with shaking at 200 rpm.
2. Pellet down 1.5 ml culture at 5000 rpm for 5 min.
3. Resuspend the pellet in 200 µl of alkaline lysis solution I. Incubate the resuspended mixture at RT for 5 min.
4. Add 200 µl of alkaline lysis solution II (alkaline SDS solution) and invert-mixed gently. Further, incubate at RT for 5 min.
5. Add 150 µl of alkaline lysis solution III (potassium acetate solution), mix it thoroughly and gently. Incubate the mix at RT for 5 min.
6. Add an equal volume (600 µl) of PCI (phenol:chloroform:isoamyl alcohol in the ratio of 25:24:1) to the tube, mix followed by centrifugation at 10,000 rpm for 10 min. Aqueous layer is collected carefully in a fresh microcentrifuge tube.

   Note: This step is optional and can be skipped.
7. Add 600 µl of C:I (chloroform:isoamylalcohol in the ratio of 24:1) and mix vigorously. Centrifuge the mixture at 10000 rpm for 10 min. The supernatant (or aqueous phase as explained in genomic DNA) thus obtained should be collected in a fresh microfuge tube.

   Note: This step is optional and can be skipped.
8. In the aqueous layer (or supernatant in case of missed 6 and 7 step), 600 µl (or 2.5 volumes) of 95 per cent ethanol (or isopropanol) should be added followed by centrifugation at 10000 rpm for 15 min.
9. Decant the supernatant and the pellet hence obtained should be washed with 70 per cent ethanol at 5000 rpm for 10 min.
10. The pelleted DNA should be air-dried and dissolved in 1 × TE containing 100 µg/ml RNase A (MQ can also be added for fast observation; RNase addition is optional).
11. Run the DNA on 1 per cent agarose gel and observe the results under UV transilluminator.

    (*Note*: Please refer to agarose gel preparation and electrophoresis (sections 4.2 and 4.3) in 'Isolation of genomic DNA' experiment)

Determination of the DNA yield can be done by

UV spectrophotometry: To lower the concentration of DNA for quantification, dilute it in TE buffer in 1:100 or 1:50 ratio (depending on copy number of plasmid) and measure the absorbance (optical density) at 260 nm and 280 nm. Use TE buffer for blank setting. The formula to calculate DNA concentration is:

$$\text{DNA } (\mu g/ml) = A_{260} \times \text{Dilution factor} \times 50$$

Here, 50 is the extinction coefficient of DNA.

The ratio of A_{260}/A_{280} estimates the purity of the DNA. DNA is pure if the ratio comes to 1.8.

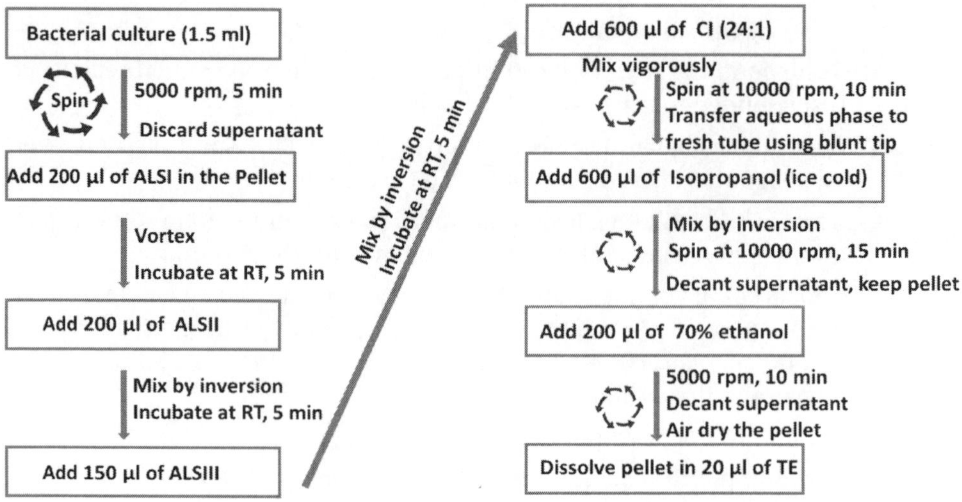

Flowchart Showing Steps of Plasmid DNA Isolation.

5. Observation and Results

The plasmid DNA bands are seen at different sizes as orange bands on the gel. The band sizes are based on mainly the conformation of DNA (Figure 2b).

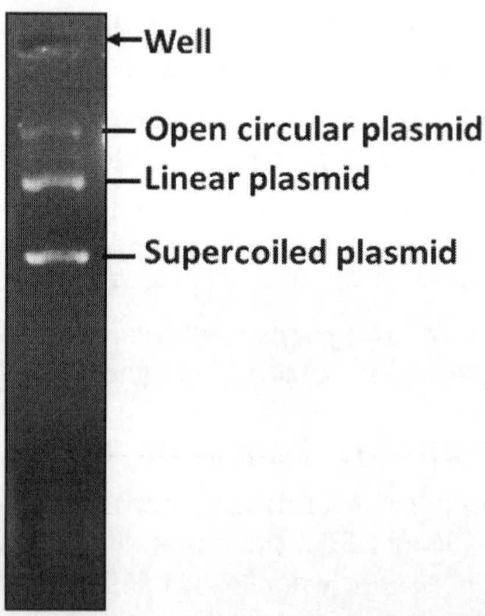

Figure 2b. Picture Showing different Forms of Plasmid.

6. Discussion

Ideally, plasmid has three types of bands that are visible on the gel, open circular, linear and supercoiled. The supercoiled DNA has a compact structure and therefore runs fastest as compared to linear and open circular. The open-circular has bulky nature due to its conformation and hence the slowest on the gel. The conformation of plasmid DNA during isolation depends on various steps followed during the isolation procedure. The harsh treatment leads to a more open circular type of plasmid DNA or in some cases might lead to a smear without any sharp bands. The size of plasmid at this stage cannot be determined by DNA ladder. DNA ladder only provides the size range and differentiates between genomic and plasmid DNA. Any contamination of genomic DNA can be monitored by running DNA ladder as plasmid DNA is short in size and run along the DNA ladder at smaller size band while genomic DNA is very large in size and therefore always visible above the DNA ladder.

7. Precautions

- ✰ All the plastic and glassware that are being used should be contamination free.
- ✰ The aqueous phase should be carefully separated from the organic phase.
- ✰ The tube content should be gently mixed after addition of ALS II.
- ✰ The pellet of DNA should be air dried thoroughly before addition of TE buffer.
- ✰ The O.D. of culture should be in the log phase for maximum yield of plasmid DNA.

8. Questions

Multiple Choice Questions

1. Cell lysis is carried out by which substance?

a. Detergents b. Water
c. Sugar solution d. Suphuric Acid

2. To isolate plasmid DNA using miniprep, bacterial cells should be harvested at

a. Lag phase b. Early log phase
c. Stationary phase d. Late log phase

3. The location of plasmid DNA can be visualized by the addition of

a. Bromophenol blue b. Ethidium bromide
c. Ortho xylene d. Texas red

4. **Which one of the following is not a form of plasmid**
   a. Linear plasmid
   b. Supercoiled Plasmid
   c. Open-circular Plasmid
   d. Coiled coil plasmid
5. **Which of the following solution is used to run the plasmid?**
   a. Agar gel
   b. Agarose gel
   c. SDS-PAGE
   d. Both a and b

Short Answer Questions

1. **What is the principle of plasmid DNA isolation?**
2. **How ALSIII helps in separation of plasmid DNA from genomic DNA?**
3. **Why antibiotic is added while culturing bacteria for plasmid isolation?**
4. **What are the various forms of plasmid DNA? Which form is visible on the gel after cutting plasmid with restriction enzyme?**
5. **What is the importance of CI during plasmid DNA isolation?**

Long Answer Questions

1. **Why do bacteria harbour plasmids naturally?**
2. **Give role of each chemical used in plasmid DNA isolation.**

REFERENCES

Maniatis T, Fritsch EF, Sambrook J (1982) Molecular cloning: A laboratory manual. Cold Spring Harbor Laboratory, Cold Spring Harbor, NY

Becker JM, Caldwell GA, Zachgo EA (2007) Biotechnology- A laboratory course. Academic Press. An imprint of Elsevier, San Diego, California.

https: //webstor.srmist.edu.in/web_assets/srm_mainsite/files/files/BT0210 per cent 20- per cent 20MOLECULAR per cent 20BIOLOGY per cent 20LABORATORYMANUAL.pdf

Sasagawa, N. Plasmid Purification. In: Gull, M., editor. Plasmid [Internet]. London: IntechOpen; 2018 [cited 2022 May 11]. Available from: https: //www.intechopen.com/chapters/60894 doi: 10.5772/intechopen.76226

Stadler, J., Lemmens, R., and Nyhammar, T. (2004). Plasmid DNA purification. *The journal ofgene medicine, 6 Suppl 1*, S54–S66. https: //doi.org/10.1002/jgm.512

3

To Perform Restriction Digestion of Plasmid DNA

Dr. Sunita K. Yadav

ABSTRACT

Restriction enzymes are natural enzymes found in bacteria that cleave DNA at specified sequences and are used to perform restriction enzyme digestion. Scientists can target a wide range of recognition sequences with hundreds of different restriction enzymes. In molecular cloning procedures such as PCR or restriction cloning, restriction enzyme digestion is widely utilised.

Keywords: *Plasmid, Fragments, Agarose gel, Linear, Supercoiled*

Learning Objectives

After performing the experiment, students will be able to:

- ☆ Explain the various types of restriction enzymes
- ☆ Perform restriction digestion of plasmid or any other DNA
- ☆ Identify different forms of plasmid DNA on agarose gel
- ☆ Analyse the result of restriction digestion.

1. Introduction

Restriction endonucleases also called "molecular scissors" are the enzymes that recognize specific sequences and cleave phosphodiester bonds of double-stranded DNA at or near that sequence without damaging it. The large fragments of DNA can be broken down into smaller fragments using restriction enzymes. These enzymes were discovered by Werner Arber, Daniel Nathans and Hamilton O. Smith, for which they received the 1978 Nobel Prize for physiology and medicine. Recombinant DNA technology is the result of restriction enzyme discovery. For the first time in the realm of Science and Medicine, restriction enzymes were used to manipulate *E. coli* bacteria that could express recombinant human insulin gene. The restriction-modification system was established in bacteria to protect them from foreign DNA, primarily bacteriophage DNA invasion. Restriction enzymes are an important aspect of the bacterial cell's restriction-modification mechanism. In general, bacteria have four restriction-modification systems that can be distinguished by restriction site cleavage specificity, subunit structure, cofactor requirement, and other factors. The specific nucleotide sequence where a restriction enzyme cuts is called restriction site which can range from 4-12 base pairs (Figure 3a). The resulting digested DNA fragments by restriction endonucleases can be resolved (separated) on agarose gel by electrophoresis. The size of the DNA fragments can be estimated by loading a DNA ladder in which size of DNA fragments are known. Based on the sequence recognition and cleaving, there are three types of restriction enzymes mainly viz: Type I, Type II and Type III. Among these restriction enzymes, Type II restriction endonucleases are the most used enzymes for genetic engineering because they cut at those sequences only that they recognize. The nomenclature of restriction enzymes is based on the organisms from which they are isolated, the strain type and total number of enzymes present. All restriction enzymes are named by abbreviating together the first letter of the genus name and first two letters of the species name followed by strain type (if any) and number of the enzyme. For example, *Hind*III enzyme name is derived from *Haemophilus influenza* strain D, 3rd enzyme. In a similar way, *Eco*RI name is based on organism *Escherichia coli*, strain

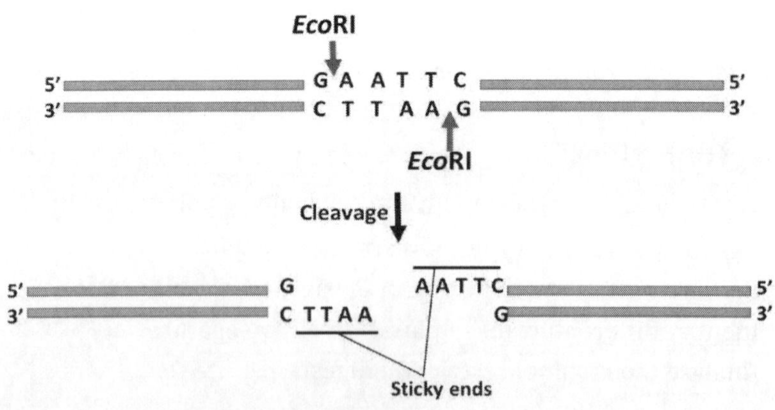

Figure 3a. Restriction Site of *Eco*RI.

RY13 and 1st enzyme. The restriction enzymes can cut DNA strands at same point on an axis resulting in blunt ends or if cut at different sites on the strands results in sticky ends.

2. Principle

The restriction endonucleases recognize 4-8 base pairs of DNA generally that are arranged in a palindromic manner. Restriction enzyme type II cuts at specific sites after recognizing them, unlike other types of restriction enzymes. The restriction enzyme requires specific reaction conditions for its activity. The enzyme buffers (also called assay buffer) contain required salt and reagents for the optimum activity of the restriction enzymes. Salt (usually NaCl) is the main constituent that varies among different restriction enzyme activity requirements. Therefore, assay buffers are precisely formulated to provide optimum salt concentration for maximal activity of enzyme. Hence, it is essential that an enzymatic reaction should always be set with an accurate buffer. While setting the reaction, it is always better to follow manufacturer's recommendations.

In the experiment, the single and double digestion of pUC18 is shown.

3. Requirements

- Enzyme: *Eco*RI, suitable assay buffer
- Chemicals: Agarose gel, electrophoresis buffer, gel loading dye, ethidium bromide, MQ water (ddH_2O)
- Plasmid: pUC18 (commercial or isolated in experiment 2)
- Equipment: Water bath (set at 37°C), centrifuge, electrophoresis apparatus
- Glassware/plasticware: Microcentrifuge tubes, micropipette, microtips, floaters

4. Procedure

1. In a 1.5 ml microcentrifuge tube, prepare the reaction mixture by adding plasmid miniprep DNA (1 μg), 10× assay buffer of *EcoRI*, and deionized water in the required volume (as written in Table 1). Add *EcoRI* enzyme at the end.
2. Mix all the contents thoroughly and pulse down the tube in a centrifuge.
3. Fix all the tubes in the floater (stand) and incubate them in waterbath (incubator) at 37°C for 1 h.
4. After 1 h, stop the reaction by adding 1 μl of 0.5 M EDTA or incubating the tubes in water bath at 65 °C for 2 min.
5. Visualise the results on agarose gel electrophoresis. Load DNA ladder in the first well to analyse the size of digested fragments.

   (*Note*: Please refer to agarose gel preparation and electrophoresis (sections 4.2 and 4.3) in 'Isolation of genomic DNA' experiment)

Table 1. The Set Up of Restriction Enzyme Digestion

Sl.No.	*Content*	*Tube 1 (Experimental/ Digested)*	*Tube 2 (Control/ Undigested Sample)*
1.	Plasmid (pUC 18)	1 μg	1 μg
2.	10× assay buffer of EcoRI	2 μl	2 μl
3.	EcoRI	1 unit (2 μl from 0.5 units/μl)	Not added
4.	Deionized water	15 μl	17 μl

5. Result and Observation

A single linearized band was seen in the lanes where restriction enzyme digested *pUC*18 was loaded. It shows that the restriction digestion was complete. The intensity of the band depends on the amount of plasmid DNA used for digestion and the activity of enzyme. The undigested *pUC*18 lane showed more than 1 band due to different forms of plasmid DNA obtained during plasmid isolation by miniprep. Lane with DNA ladder shows many bands corresponding to size of the DNA (Figure 3b).

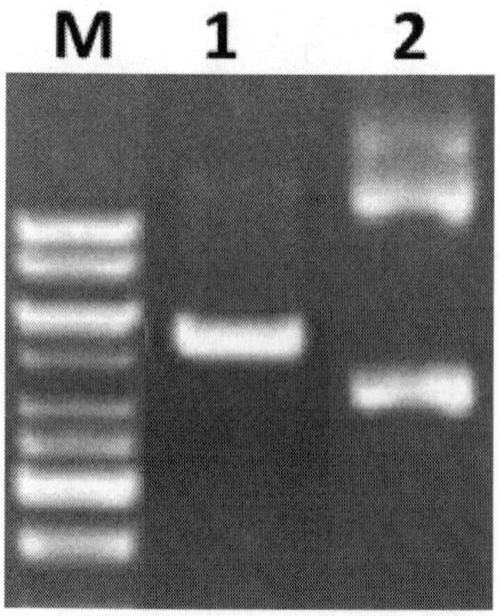

Figure 3b. Depiction of pUC18 Digestion.
Lane M contains DNA marker. Lane 1 shows digested pUC18 with *EcoRI* while lane 2 depicts undigested pUC18 showing different forms of plasmid DNA.

6. Applications

- Restriction enzymes are used in molecular cloning to produce recombinant proteins in large quantity for various purposes *i.e.* as drugs, hormones, vaccines or as part of a pathogen detection kit.
- Restriction enzymes are also used for production of DNA library. In DNA library, isolated DNA fragments (fragments produced after restriction enzyme digestion) are cloned in suitable vectors to make a collection of

whole genome or mRNA expressing genes of an organism (called genomic DNA library or cDNA library respectively).

- Restriction enzymes are used for cloning sequences that are subjected to site-directed mutagenesis. This is useful to find out the basis of genetic diseases and mapping the active site of the protein
- Restriction enzymes are used for restriction mapping. The restriction enzymes digest a piece of DNA into fragments and then overlapping sequences of the fragments help in aligning the sequences in the order.

7. Precautions

- All the plastic and glassware should be sterile.
- Plasmid DNA should be quantified before setting the reaction
- Pipetting of the required volume should be done properly
- Temperature of water-bath should be maintained throughout the experiment
- All the contents of reaction mixture should be thoroughly mixed
- Wear gloves while working with ethidium bromide.
- Optimum concentration of restriction enzyme should be used (high concentration of enzyme leads to star activity)

  (*Note*: Star activity is the non-specific cleavage of DNA by restriction enzymes in the presence of altered environmental conditions such as high salt concentration, impurities presence or high amount of enzyme present as compared to DNA. It results in loss of specificity and enzymes cleave DNA sequences other than their specific sites)
- All the reagents including buffers and enzymes should be stored at the required temperature before and after setting the reaction.

8. Questions

Multiple Choice Questions

1. **The most probable site for restriction enzyme would be which of the following sequences in double-stranded DNA?**

   a. AAGG/TTCC b. AGTC/TCAG
   c. GGCC/CCGG d. ACCA/TGGT

2. **Which of the following restriction enzymes are used mostly in recombinant DNA technology?**

   a. Type I b. Type II
   c. Type III d. Type IV

3. **The size of linearized DNA can be estimated by using:**
   a. DNA scale
   b. DNA cutter
   c. DNA marker
   d. DNA stainer
4. **The restriction enzymes occur naturally in bacteria as part of :**
   a. Restriction-manipulation system
   b. Restriction-endonuclease system
   c. Restriction-nucleation system
   d. Restriction-modification system
5. **The specific nucleotide sequence where a restriction enzyme cuts is called restriction site which can range from:**
   a. 4-12 base pairs
   b. 2-20 base pairs
   c. 4-30 base pairs
   d. 6-8 base pairs

Short Answer Questions

1. **Why the restriction enzymes are called 'molecular scissors?**
2. **Which form of plasmid is seen on gel after cutting it with restriction enzyme?**
3. **How do you measure the size of plasmid on gel?**
4. **What are multiple cloning sites in plasmid?**

Long Answer Questions

1. **What are the characteristics of a good plasmid?**
2. **Give functions of naturally occurring plasmids.**
3. **What are the different types of restriction enzymes?**

REFERENCES

Becker JM, Caldwell GA, Zachgo EA (2007) Biotechnology- A laboratory course. Academic Press. An imprint of Elsevier, San Diego, California.

https: //webstor.srmist.edu.in/web_assets/srm_mainsite/files/files/BT0210 per cent 20- per cent 20MOLECULAR per cent 20BIOLOGY per cent 20LABORATORYMANUAL.pdf

Maniatis T, Fritsch EF, Sambrook J (1982) Molecular cloning: A laboratory manual. Cold Spring Harbor Laboratory, Cold Spring Harbor, NY

4

Preparation of Competent Cells and Transformation of the Cells Using pUC18 to Calculate the Transformation Efficiency

Dr. Sunita K. Yadav

ABSTRACT

Heat-shock transformation using calcium chloride is a strong molecular biology approach for introducing foreign DNA into a host cell. The idea behind the procedure is to use $CaCl_2$ to make cells competent for plasmid introduction. Plasmids often contain the desired gene(s) as well as markers for selection such as antibiotic resistance gene. Chemically competent cells are generated using the calcium chloride approach. The uptake of plasmid by a competent cell is aided by heat shock. Molecular biology applications, such as amplification of plasmid or expression of recombinant protein, will be possible with transformed cells. The methodology for preparing calcium competent E. coli utilising the heat-shock transformation technique to uptake the plasmid from the environment is outlined in the chapter.

Keywords: *Heat-shock, Competent cell, Calcium chloride, Plasmid, Transformation.*

Learning Objectives

After the practical, students will be able to:

- ✰ Prepare competent cells for transformation and understand the concept behind the usage of competent cells
- ✰ Transform the cells with a foreign plasmid
- ✰ Spread the transformed bacteria on a plate
- ✰ Calculate transformation efficiency
- ✰ Apply the knowledge for other applications such as cloning, recombinant protein expression, *etc.*

1. Introduction

Bacterial transformation is the process of genetic modification of a cell from direct uptake (horizontal gene transfer) and incorporation of foreign genetic material from its surroundings through the cell membrane. It was first reported in 1928 by Griffith in *Streptococcus pneumoniae* followed by Avery *et al. (*1944) demonstrating the transforming principle of DNA. The transformation term has been used differently for different types of cells. When the DNA is introduced with the help of viruses, it is called transduction; if it happens through cell to cell contact such as in bacteria, it is denoted as conjugation while the transformation of animal cells is termed transfection.

In the transformation process, the transfer of gene does not require a living cell as donor of gene, rather the presence of DNA in the environment is sufficient to be taken by the cell. The bacteria should have the ability to take foreign DNA efficiently, such bacteria are called competent cells. The competence among bacteria is decided by several factors and it differs with genus. The entered foreign DNA can be degraded by nucleases and if the genetic material is comparable to the host DNA, it can be integrated in to the host genome. Occasionally, the introduced foreign DNA remained as plasmid co-existing with chromosomal DNA.

In the transformation process, the introduction of DNA into bacterial cell sis carried out in the early logarithmic phase when cells are actively dividing. Since DNA is highly hydrophobic, it does not normally pass through bacterial cell membrane. Therefore, to carry out transformation, non-competent cell must be made competent to take up foreign DNA. One of the methods to make competent cells is using calcium chloride ($CaCl_2$) method. The competent cells are then used for transformation via chemical method or heat-shock method or electroporation.

2. Principle

The heat-shock transformation promotes $CaCl_2$-treated bacterial cells to take up DNA from their surroundings. The specific mechanism behind its working is still not certain, however many components of the procedure have been hypothesised.

Calcium ions are thought to act as a cation bridge between the negative charges on phosphorylated lipid A in lipopolysaccharide (LPS) surface of the bacterial cell and the phosphate ion backbone of DNA present in the suspension. The chilled $CaCl_2$ solution aids DNA attachment to the cell surface, which allows DNA to enter the cell after a brief heatshock phase. Selection markers such as antibiotic resistance or fluorescence are commonly used to identify effectively transformed cells. This approach is extensively used to transform cells with plasmids for a variety of applications, including expression of recombinant protein, cloning, and long-term plasmid storage.

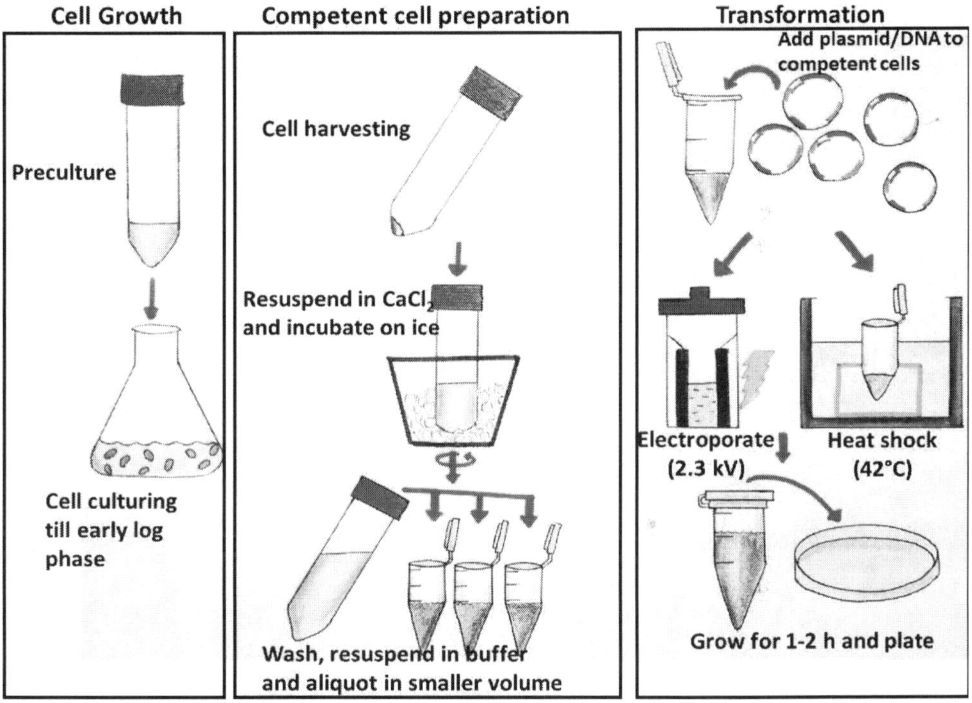

Figure 4a. Steps of Transformation Process (Illustrator: Aakriti Singh, Department of Botany, University of Delhi).

3. Requirements

LB agar plate with bacterial colonies, LB media in flask, LB agar petriplate, $CaCl_2$, Glycerol, Spreader, Spirit lamp, Temperature regulated waterbath, Laminar air flow hood with UV lamp, Incubator shaker, Wire loop, Tips, Micropipette, Microtube stand, Microcentrifuge tubes, Ice.

4. Procedure

4.1. Preparation of Competent Cells (Figure 4a)

[*Note*: Competent *E. coli* cells (DH5α) preparation method described here is a minor modification of the method described by Cohen *et al.* (1972)].

1. For primary culture, inoculate a single colony of the bacteria (*E. coli* or desired strain) from the streaked/plated petriplate in 5 ml LB media at 37 °C for O/N at 200 rpm.
2. Inoculate LB medium (100 ml) with 1 per cent primary culture and grow it till absorbance at 600 nm reaches 0.3–0.4.
3. Chill the obtained culture on ice for 30 min and harvest the cells by centrifugation at 5,000 rpm for 10 min.
4. Resuspend the obtained cell pellet in 30 ml of chilled 100 mM $CaCl_2$ following an incubation on ice for 1 h followed by centrifugation at 5,000 rpm for 10 min at 4 °C.
5. Again resuspend the obtained pellet in 2 ml of 100 mM $CaCl_2$ followed by incubation at 4°C for O/N.
6. Add glycerol at a concentration of 20 per cent and mix gently.
7. Make small aliquots of 100/200 µl each and store at –80 °C till further use.

4.2. Transformation of Competent Cells (Figure 4a)

[*Note*: The competent *E. coli* cells are transformed with the ligation mix or with plasmid DNA as described by Maniatis *et al.* (1982)]

1. Thaw the competent cells on ice that are taken out from –80 °C.
2. Mix gently the supercoiled DNA (pUC18 or the ligation mix) with 100/200 µl of competent cells and incubate on ice for 30 min.
3. Give heat shock to the cells at 42 °C for 90 sec with an immediate incubation on ice for 5 min.
4. Add LB (~900 µl) to the cells and grow it for 1 h at 37 °C at 200 rpm.
5. Harvest the cells by centrifugation at 5,000 rpm and resuspend in 30-80 µl of LB.
6. Spread the resuspended cells by plating them on LB agar plate (containing suitable antibiotic/s) and incubate for O/N at 37 °C.

4.3. Calculation of Transformation Efficiency

The transformation efficiency is defined as the number of transformants generated per µg of supercoiled plasmid DNA used in the transformation reaction.

1. Count the generated colonies on the plate or in the given data to determine colony-forming units (CFU) (Figure 4b)

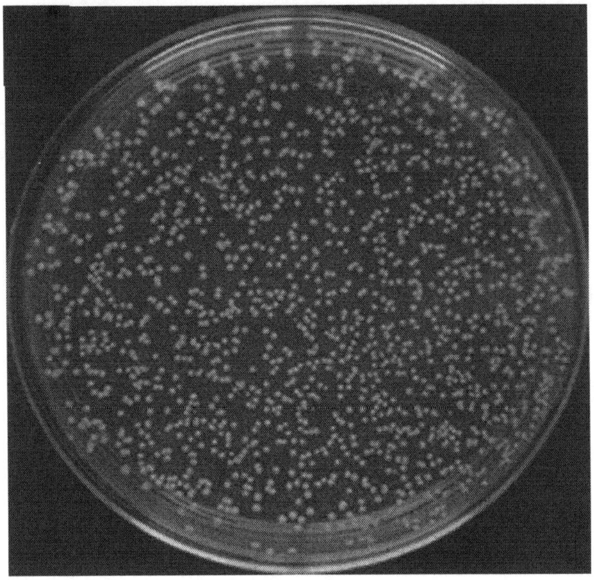

Figure 4b. A Picture Showing Isolated Single Colonies of Bacteria on Petriplate.

2. Determine the amount of plasmid DNA (in μg) used to spread on the plate.
3. Divide the number of colonies (in CFU) on the plate with the amount of DNA used to spread on the plate (in μg).
4. The results should be expressed in CFU/μg.

Transformation efficiency is calculated using the formula below:

$$\text{Transformation efficiency} = \frac{\text{No. of transformants}}{\text{Amount of DNA}(\mu g)} \times \frac{\text{Final volume at recovery (ml)}}{\text{Vol. plated (ml)}}$$

5. Calculation

Example: 100 ng of supercoiled DNA is mixed with 100 μl of competent cells. 900 μl of LB media added for revival and 200 μl of this volume is used to spread on the plate. Assume 80 colonies are observed on the plate. Calculate transformation efficiency of the competent cells used in the experiment.

Ans

Step 1: No. of colonies (CFU): 80 (given)

Step 2: Amount of supercoiled DNA (μg) spread on the plate:

DNA spread on plate:

$$\frac{\text{Volume spread } (\mu l) \times \text{Total DNA used for transformation } (\mu g)}{\text{Total volume: Volume of competent cell + Volume used for revival of bacterial}}$$

$$= \frac{200\mu l \times 0.1\mu g}{1000\mu l} = 0.02\mu g$$

Step 3: Transformation efficiency: divide the no. of colonies (CFU) on the plate by the amount of DNA (µg) used to spread on the plate

= 80/0.02 = 4000 CFU/µg of DNA = 0.4×10^4 CFU/µg DNA

Answer: 0.4×10^4 CFU/µg DNA

6. Discussion

There are several factors that affect transformation efficiency. Few of them are discussed below.

a. **Type and size of DNA:** Transformation of supercoiled DNA is more efficient as compared to linear and ssDNA. The relaxed plasmids transformed at around 75 per cent as compared to the total efficiency of supercoiled DNA. Single-stranded DNA transformed at much lower efficiency than dsDNA that is 10^4 lower efficiencies. The efficiency of transformation regresses with increasing size of DNA *i.e.*, smaller plasmids/DNA transform more competently than larger plasmids/DNA.

b. **Type of bacteria strain:** Some bacterial strains are more efficiently transformed than others. Since, the cell wall thickness and composition of the outer cell wall of Gram-positive and Gram-negative bacteria are different, their efficiency to be transformed with foreign DNA also differs. The Gram negative bacteria are easy to transform as compared to Gram-positive.

c. **Amount of salt:** The salts present in the transformation mixture may decrease the efficiency. Therefore, volume of plasmid DNA can be limited to 1 µl per transformation reaction. Also, high quality purified DNA should be used if possible that are free from salts and other contaminants.

d. **Ligation mixture:** Since, ligation mixture contains ligases, the transformation of cells becomes difficult as ligases inhibit electroporation. Therefore, ligases should be heat-inactivated at 65°C for 5 minutes before adding to the cells.

e. **Freeze/thawing of competent cells:** The freeze/thaw cycle of competent cells decreases the efficiency as much as two-fold. Hence, small aliquots of competent cells should be prepared and a whole vial should be used at a time.

f. **Time between transformation reaction and plating:** The transformation efficiency is inversely proportional to time difference between transformation and plating. However, it depends on few other parameters also such as the type of competent cell strain used, type of DNA and type of method used for transformation.

7. Applications of Transformation

Transformation method is extensively used for various molecular biology applications. The transformed bacteria (an inserted foreign gene) is generally used as a host organism for several applications such as:

- ☆ Yield of DNA can be increased by multiplication of DNA in the host organism and a higher concentration of DNA can be purified.
- ☆ The isolated and purified DNA from the transformed host organism can be used for various cloning procedures.
- ☆ The transformed host can be used to express large amounts of proteins and enzymes from inserted DNA.
- ☆ The transformed cells can be used for the generation of cDNA libraries.
- ☆ The DNA can be used for linkage and recombination studies in the transformed host.

8. Precautions

- ☆ The experiment should be done in sterilized environment (inside UV hood)
- ☆ Heat shock temperature as well as the timing should be properly maintained.
- ☆ The colony should be properly counted keeping dilution factor in mind.
- ☆ The agar plate should be moisture free before plating the culture.
- ☆ The plate should be kept upside down overnight.

Few more Points to Remember

- ☆ Supercoiled DNA should be used to calculate transformation efficiency of competent cells.
- ☆ The most efficient transformation results are obtained when <10 ng of DNA is used. The higher amount of DNA becomes a limiting factor for the transformation.

9. Questions

Multiple Choice Questions

1. **The process of taking foreign gene by mammalian cells is called:**
   a. Transduction b. Transfection
   c. Transformation d. Transgene
2. **If a virus is involved in introducing a foreign gene, the process is termed as:**
   a. Transduction b. Transfection
   c. Transformation d. Transgene
3. **The best transformation efficiency is observed with:**
   a. Single stranded DNA b. Supercoiled DNA
   c. Double stranded DNA d. Supercoiled DNA
4. **The transformation efficiency is defined in:**
   a. CFU/μg b. CFU/ng
   c. Colonies/meter d. Colonies/megabyte
5. **The best method to transform plant cell is:**
   a. Transfection b. Heat shock
   c. Electroporation d. None of these

Short Answer Questions

1. **Define transformation.**
2. **What should be the density range of grown bacteria to be used for competent cells preparation?**
3. **Why LB media with antibiotic is used for revival of bacteria?**
4. **Describe the role of $CaCl_2$ in transformation?**
5. **Why competent cells are prepared to calculate transformation efficiency?**

Long Answer Questions

1. **A student has added 5 μl of 65 ng/ml pGEMT plasmid to a tube of *E.coli* DH5α *(*bacteria). The bacteria were suspended in 80 μl of chilled $CaCl_2$. Post heat shock, 400 μl of LB medium was added to the treated tube. 200 μl of cell suspension was spread on agar plates with ampicillin. The plates were incubated for 22 hours in incubator at 37°C. 150 colonies were counted on the spread plate next day.**

   Answer the following questions:

   a) How much plasmid (PGEMT) was added to competent bacteria?

b) How much plasmid was taken for spreading on plate?

c) Calculate Transformation Efficiency.

b) Explain different factors that affect transformation efficiency.

REFERENCES

A decomposable silica-based antibacterial coating for percutaneous titanium implant - Scientific Figure on ResearchGate. Available from: https: //www.researchgate.net/figure/Bacteria-colonies-cultivated-on-Mueller-hinton-agar-plates-from-A-s-Ti-B-MaO-Ti-and_fig4_312409530 [accessed 15 May, 2022]

Bergmans, HE, van Die,I.M., Hoekstra, WP. 1981. Transformation in *Escherichia coli*: stages in the process. J. Bacteriol. 146: 564-570.

Chang, Angela Y., Chau, Vivian WY., Landas, Julius A., Pang, Yvonne. 2017. Preparation of calcium competent *Escherichia coli* and heat-shock transformation. JEMI methods, vol 1: 22-25

Cohen SN, Chang AC, Hsu L. Nonchromosomal antibiotic resistance in bacteria: genetic transformation of *Escherichia coli* by R-factor DNA. Proc Natl Acad Sci U S A. 1972 Aug;69(8): 2110-4. doi: 10.1073/pnas.69.8.2110. PMID: 4559594; PMCID: PMC426879.

http: //bcas.du.ac.in/wp-content/uploads/2020/04/ACB-TRANSFORMATION-EFFICIENCY-PRACTICAL.pdf

Maniatis T, Fritsch EF, Sambrook J (1982) Molecular cloning: A laboratory manual. Cold Spring Harbor Laboratory, Cold Spring Harbor, NY

5

Construction of Circular and Linear Restriction Maps from the Data Provided

Dr. Sunita K. Yadav

ABSTRACT

Restriction mapping is a technique for mapping the relative position of cleavage site for restriction enzyme in an unknown section of DNA. It can be achieved by breaking it down into pieces using restriction enzymes and then determining where the cleavage sites are located. This approach relies on restriction endonucleases, which are enzymes that can cut or digest DNA molecules at short, particular sequences known as restriction sites. The fragments of restriction enzyme digested DNA can be analysed using gel electrophoresis, a laboratory technique for separating pieces of DNA based on their size. After assessing the various fragments separated based on their sizes, they can be aligned in a sequence and hence a restriction map can be created.

Keywords: *Fragment, Restriction site, Circular map, Linear map.*

Learning Objectives

After the experiment, students will be able to:

- ☆ Understand the importance of restriction mapping.
- ☆ Comprehend the concept of double digestion in restriction mapping.
- ☆ Interpret the sizes of separated DNA fragments on an agarose gel.
- ☆ Solve the problems related to linear and circular restriction map construction.

1. Introduction

Restriction mapping is a process that is used to map an unidentified fragment of DNA when it is digested in to pieces by restriction enzymes (also called restriction endonucleases). Different types of restriction enzymes can be employed for restriction enzyme digestion of DNA molecules. These enzymes cut the DNA fragment in short sequences that are specifically cut on specific location recognised by a restriction enzyme. The resulting fragment upon digestion with restriction endonucleases can be screened for size by running them on DNA gel electrophoresis. The gel electrophoresis resolves the DNA fragments according to their sizes.

Restriction Enzymes

Restriction enzymes (Type II) are endonucleases that recognizes the DNA at a specific sequence and cut at those sites. These restriction enzymes are part of the restriction-modification system in bacteria and help in defense mechanism against bacteriophages. The enzymes part of modification system helps in modifying the bacterial genome by adding methyl group to protect it from cleaving by endonucleases. The restriction enzyme sites are palindromic that is they read alike from both sides. There are several restriction enzymes that have been isolated from various bacteria that recognize different restriction sites on DNA strand. The DNA fragment digested with same restriction endonuclease give same results everytime; therefore, the reproducibility of the result is very high. Nowadays, restriction enzymes are commercially available and they are inseparable part of genetic engineering.

2. Principle of Restriction Mapping

The most common method for the construction of a restriction map of DNA sample is digesting it with restriction enzymes. The DNA sample is divided in to three parts and among these, two parts should be digested separately with two restriction enzymes individually. The third part of sample should be digested with both restriction enzymes at the same time. The digested fragments need to be separated using gel electrophoresis. Total size of the fragments in each digestion should be equal. Since the digested fragment size depends on the location of restriction enzyme sites, each restriction enzyme site can be drawn on the map according to the fragment length. The double digestion of fragment is chiefly valuable for appropriately mapping the sites. The restriction map is the final depiction of the DNA fragment showing the positions of restriction enzyme sites.

3. Problems

Construct a restriction map using the following data. Write the size and position of restriction sites and draw the electrophoretic gel picture.

Problem 1

No. of Fragments	*Name of Enzyme*	*Fragments Size*
1	*EcoRI*	25kb
4	*BamHI*	13 kb, 6 kb, 4 kb, 2 kb
5	*EcoRI+BamHI*	7 kb, 6 kb (2), 4 kb, 2 kb

Solution

Since sum of no. of fragments cut with a single enzyme is equal to the no. of fragments observed after double digestion, the restriction map is circular.

- ✰ As only one fragment (25 kb) is present with *EcoRI*, place one site of *EcoRI* on circular map anywhere.
- ✰ The 13 kb fragment of *BamHI* is present as 7 kb and 6 kb fragment in double digestion, place 7 kb and 6 kb on the circular map with *BamHI* site at 5' end of 7 kb and 3' end of 6 kb.
- ✰ Since 6 kb fragment is present in *BamHI* and double digestion, place *EcoRI* site at the junction of 7 kb and 6 kb fragments of step 2 and mark 1 more 6 kb fragment adjacent to 6 kb fragment of step 2 with *BamHI* site at its 3' end.
- ✰ The two remaining fragments, 4 kb and 2 kb are present in *BamHI* as well as in double digestion, mark both of them with *BamHI* sites at the ends adjacent to 6 kb fragment or 7 kb fragment.

Hence, the electrophoretic picture and restriction map are shown in Figure 5a.

Problem 2.

No. of Fragments	*Name of Enzyme*	*Fragments Size*
2	*HindIII*	5.3 kb, 1.5 kb
2	*NotI*	4.1 kb, 2.7 kb
4	*HindIII +NotI*	3 kb, 2.3 kb, 1.1 kb, 0.4 kb

Solution

Since sum of no. of fragments cut with a single enzyme is equal to the no. of fragments observed after double digestion, the restriction map is circular.

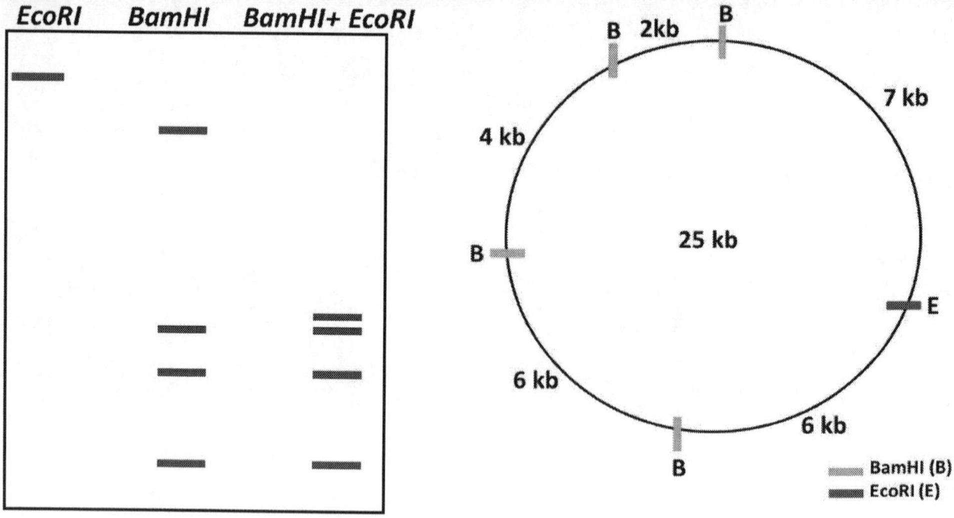

Figure 5a

- ☆ Assign alphabetic orders to the fragments of *HindIII* and *NotI i.e.* a=3, b=2.3, c= 1.1, d=0.4.
- ☆ In *HindIII* fragments, 5.3 kb=a+b and 1.5 kb= c+d
- ☆ In *NotI* fragments, 4.1 kb= a+c and 2.7 kb= b+d
- ☆ Therefore, the order would be c-a-b-d.
- ☆ Arrange them in circular map in the above order.

Hence, the electrophoretic picture and restriction map are shown in Figure 5b.

Problem 3

No. of Fragments	*Name of Enzyme*	*Fragments Size*
3	*BamHI*	9 kb, 3 kb, 0.5 kb
2	*SalI*	7.5 kb, 5 kb
4	*BamHI +SalI*	7 kb, 3 kb, 0.5 kb, 2 kb

Solution

Since sum of no. of fragments cut with a single enzyme is more than the no. of fragments observed after double digestion, the restriction map is linear.

- ☆ Total size of the linear map is 9 + 3 + 0.5 kb. *i.e.* 12.5 kb
- ☆ The largest fragment of *BamHI* is 9 kb. Place it first.
- ☆ Now 9 kb exists as 7 + 2 kb in double digestion (*BamHI+SalI*), so place 7 and 2 kb on 9 kb linear map.

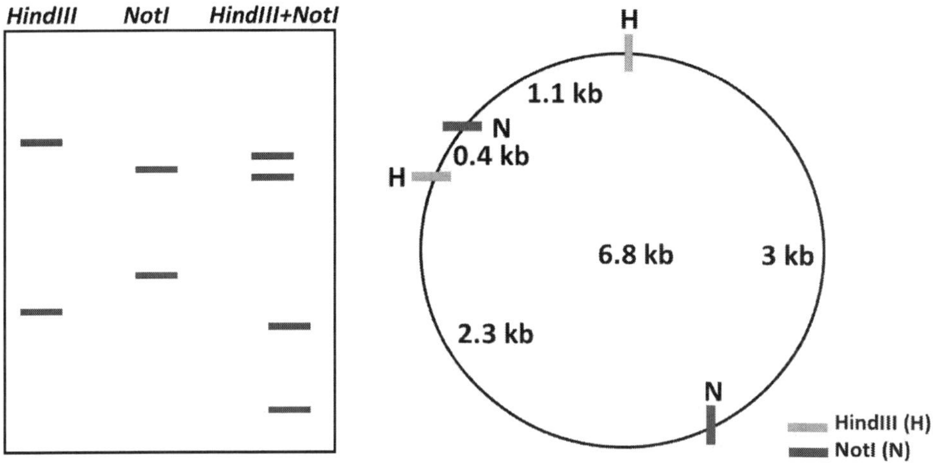

Figure 5b

- ✰ Now the largest fragment of *SalI i.e.* 7.5 kb exists as 7 and 0.5 kb in double digestion, place 0.5 kb adjacent to 7 kb.
- ✰ The 5 kb of *SalI* exists as 3 kb and 2 kb in double digestion, hence place 3 kb adjacent to 2 kb on linear map.
- ✰ Therefore the sequence of double digestion is 0.5 kb, 7 kb, 2 kb, 3 kb.

Hence, the electrophoretic picture and restriction map are shown in Figure 5c.

Problem 4

No. of Fragments	*Name of Enzyme*	*Fragments Size*
2	*NotI*	2.6 kb, 1.1 kb
3	*HindIII*	1.6 kb, 0.8 kb, 1.3 kb
4	*NotI + HindIII*	1.6 kb, 1 kb, 0.8 kb, 0.3 kb

Solution

Since sum of no. of fragments cut with a single enzyme is more than the no. of fragments observed after double digestion, the restriction map is linear.

- ✰ Total size of the linear map is 2.6 + 1.1 kb *i.e.* 3.7 kb
- ✰ The largest fragment of *NotI* is 2.6 kb. Place it first.
- ✰ Now 2.6 kb exists as 1.6 + 1 kb in double digestion (*NotI+HindIII*), so place 1.6 and 1 kb on 2.6 kb linear map.
- ✰ Now 1.6 kb and 0.8 kb of *HindIII* remain same after double digestion also. But 1.3 kb of *HindIII* exists as 1 kb + 0.3 kb on double digestion. Therefore place 0.3 kb near 1 kb of step 3.

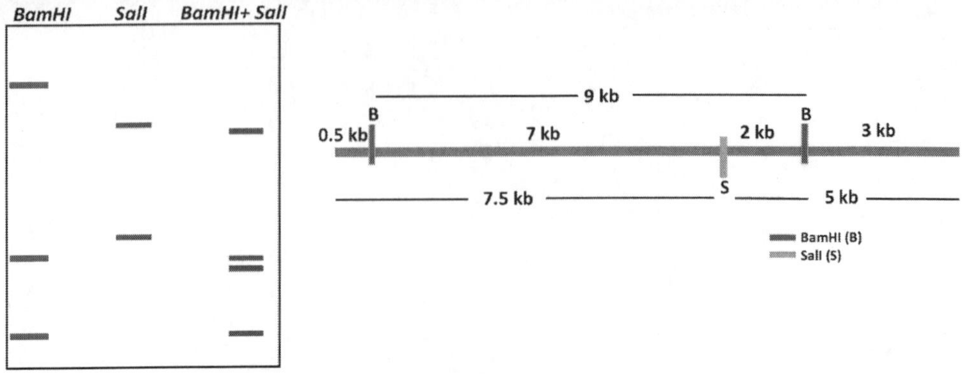

Figure 5c

- ☆ The 1.1 kb of *NotI* exists as 0.8 +0.3 kb in double digestion, place 0.8 kb near 0.3 kb on linear map.
- ☆ Therefore the sequence of double digestion is 1.6 kb, 1 kb, 0.3 kb, 0.8 kb.

Hence, the electrophoretic picture and restriction map are shown in Figure 5d.

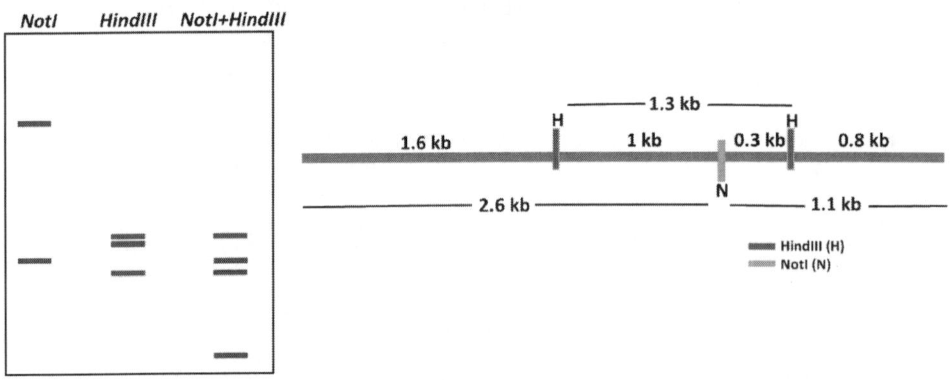

Figure 5d

4. Applications

- ☆ Detection of genetic defects.
- ☆ DNA fingerprinting.
- ☆ Mapping of the human genome.

5. Disadvantages

- ☆ Restriction enzyme mapping does not reveal the correct orientation of the restriction sites; however, the relative position of restriction enzyme sites remains same.

- ☆ In the case of insufficient enzyme or less incubation time, the plasmid is not digested properly and hence the results of restriction mapping may not be clear and confusing.
- ☆ It is generally applicable for smaller molecules than larger ones.

6. Questions

1. Construct a restriction map using the following data. Write the size and position of restriction sites and draw the electrophoretic gel picture.

A.

No. of Sites	*Name of Enzyme*	*Fragments Size*
1	*HindIII*	2.5 kb, 0.9 kb
2	*BamHI*	1.5 kb, 0.8 kb, 1.1 kb
3	*BamHI + HindIII*	1.5 kb, 1 kb, 0.8 kb, 0.1 kb

B.

No. of Sites	*Name of Enzyme*	*Fragments Size*
2	*EcoRI*	9 kb, 6 kb, 3 kb
1	*SalI*	11 kb, 7 kb
3	*EcoRI + SalI*	9 kb, 6 kb, 2 kb, 1 kb

C

No. of Sites	*Name of Enzyme*	*Fragments Size*
2	*HindIII*	3.5 kb (2)
3	*EcoRI*	4 kb, 2 kb, 1 kb
5	*EcoRI + HindIII*	2 kb, 2.8 kb, 0.7 kb, 1.2 kb, 0.3 kb

REFERENCES

Gene Cloning and Analysis, T A Brown, 7th Edition

https: //www.bioted.es/protocolos/ED-206-ENG.pdf

Maniatis, T, Fritsch, E F and Sambrook, J (1982) Molecular Cloning. A Laboratory Manual, Cold Spring Harbor Laboratory

Primrose S B, R M Twyman and R W Old. Principles of Gene Manipulation. 6th edn.

Smith, H. O., and Birnstiel, M. L. (1976). A simple method for DNA restriction site mapping. *Nucleic acids research, 3*(9), 2387–2398. https: //doi.org/10.1093/nar/3.9.2387

6

To Study the Technique of DNA Fingerprinting

Dr. Sonika Sharma

ABSTRACT

DNA Fingerprinting is a technique to identify individuals. It makes use of the unique regions of DNA, the satellite DNA to make a unique pattern called the DNA Fingerprint. This technique was discovered serendipitously by Dr. Alec Jeffreys at the University of Leicester, UK in 1984. DNA Fingerprinting is a technique routinely used in forensics for identifying individuals. This technique relies on the unique repetitive regions of DNA and is highly accurate and reliable in identifying individuals. Initially the VNTR regions were used to make the DNA Fingerprint, but the technique was tedious and cumbersome. Later with the discovery of Polymerase chain reaction and its application to DNA Fingerprinting, the technique became quick and easy. It is extensively used in identifying individuals in paternity suits, crime scenes and military establishments. All aspects of the technique related to principle, procedure and applications will be addressed in this chapter.

Keywords: *DNA Fingerprint, VNTRs, Forensics, RFLP*

Learning Objectives

After the practical, students will be able to:

- ✰ Understand RFLP
- ✰ Understand the principle of DNA Fingerprinting
- ✰ Differentiate between Multilocus DNA Fingerprints and Single Locus DNA Fingerprints
- ✰ Understand, compare and conclude from DNA fingerprints provided

1. Introduction

Fingerprinting is a method of identifying individuals based on the pattern of ridges on their fingers as these patterns are unique to every individual. With the massive expansion of various molecular biology tools and techniques, the use of DNA to create a unique pattern was conceptualized and is known as DNA fingerprinting/DNA profiling. Invented by Sir Alec Jeffreys in 1984 (Figure 6a), it has been extensively used by scientists to distinguish between individuals of the same species using their DNA.

Figure 6a. Sir Alec Jefferys.

All individuals have 99.9 per cent of their ~ 3 million nucleotides genome identical, as the coding sequences cannot be altered. So, the unique region comprising 0.1 per cent of the genome is utilized to build a unique DNA fingerprint. These variable regions are present in the heterochromatin and are called satellite DNA. Satellite DNA consists of arrays of tandem repeats (arranged head to tail) and the repeat units vary between individuals which gives it its uniqueness. The three types of satellites are

1. Microsatellites/Simple Sequence repeats (SSRs) are small arrays of tandem repeats of 1-7 bases.
2. Minisatellites/Variable number of Tandem repeats (VNTRs) have tandem repeats of 6-100 bases (Figure 6b).
3. Macrosatellites have a core sequence ranging from 100 to 6500 bp.

Initially, the minisatellites (VNTRs) were utilized to make the unique pattern *i.e.*, the DNA Fingerprint.

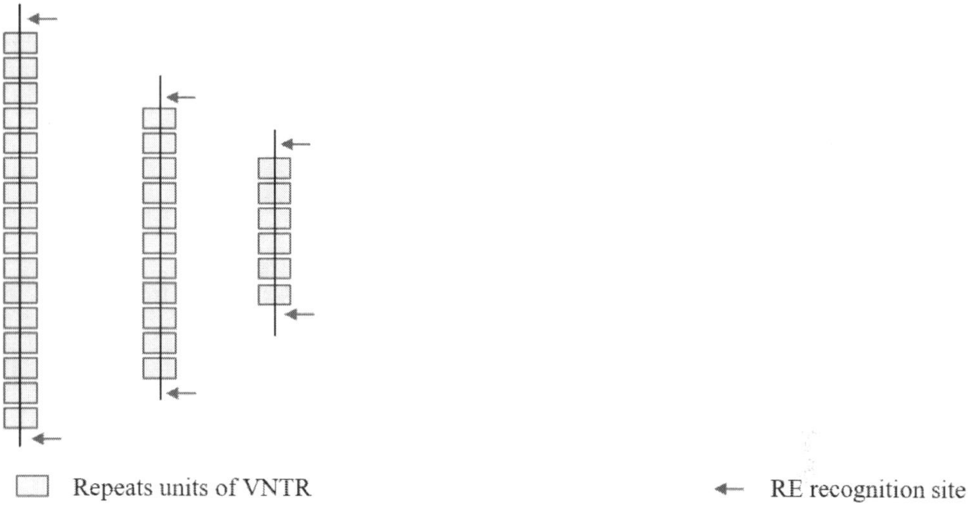

Figure 6b. Variable Number of Tandem Repeats (VNTRs).

2. Principle

The technique requires DNA from the individual whose DNA fingerprint has to be made. DNA can be obtained from blood, buccal cells, or hair follicles. The DNA thus extracted is subjected to restriction digestion. The satellite regions are usually flanked by restriction enzyme sites and hence the difference in the number of tandem repeats between individuals can be mapped using this technique called Restriction Fragment Length Polymorphism (RFLP). In principle, RFLP is a technique that detects polymorphisms arising due to differences in restriction enzyme recognition sites (Figure 6c).

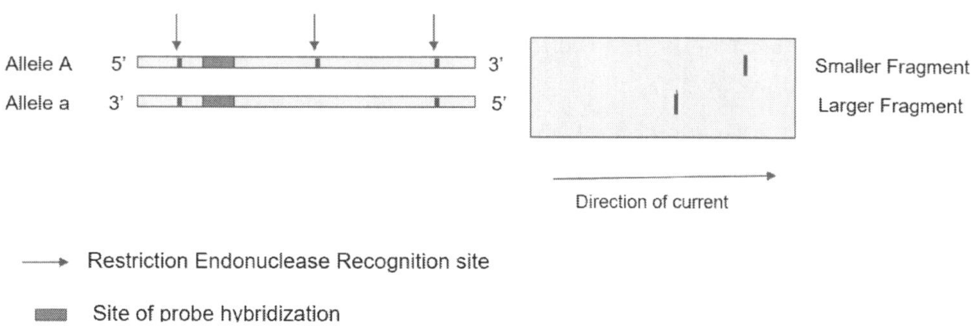

Figure 6c. Restriction Fragment Length Polymorphism (RFLP).

The next step is to separate out the fragments thus generated on an agarose gel and subsequently transfer these bands to a nylon membrane. To detect the unique regions of the satellite DNA we make use of probes. Some probes are designed to hybridize to many satellite regions in the genome and are thus called multilocus probes. On the other hand, Single locus probes hybridize specifically to a particular satellite region. These probes are radioactively labelled and their hybridization can be detected by another technique of autoradiography. The result of probing is a unique pattern showing many bands in case of multilocus probes, and a maximum of two bands in the case of single locus probing. The pattern thus obtained is the unique DNA fingerprint (Figure 6d).

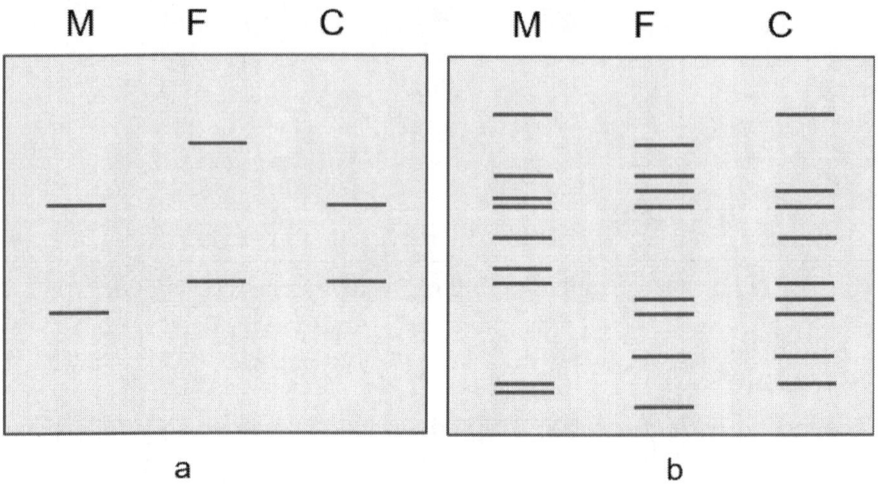

Figure 6d. A Single Locus Probe (a) Hybridizes to and detects a maximum of 2 bands as compared to a Multilocus probe (b) Which hybridizes and detects many bands. (M- mother; F- father; C- child).

3. Procedure

The procedure can be summarized as follows (Figure 6e):

1. Extraction of DNA
2. Restriction Digestion of DNA to create a RFLP
3. Electrophoretic separation of fragments on Agarose gel
4. Transfer of the fragments on nylon/Nitrocellulose membrane (Southern Hybridization)
5. Hybridization with labelled probe either Single locus or multilocus and washing off of unbound probe
6. Detection of probe hybridization by autoradiography to obtain DNA fingerprint

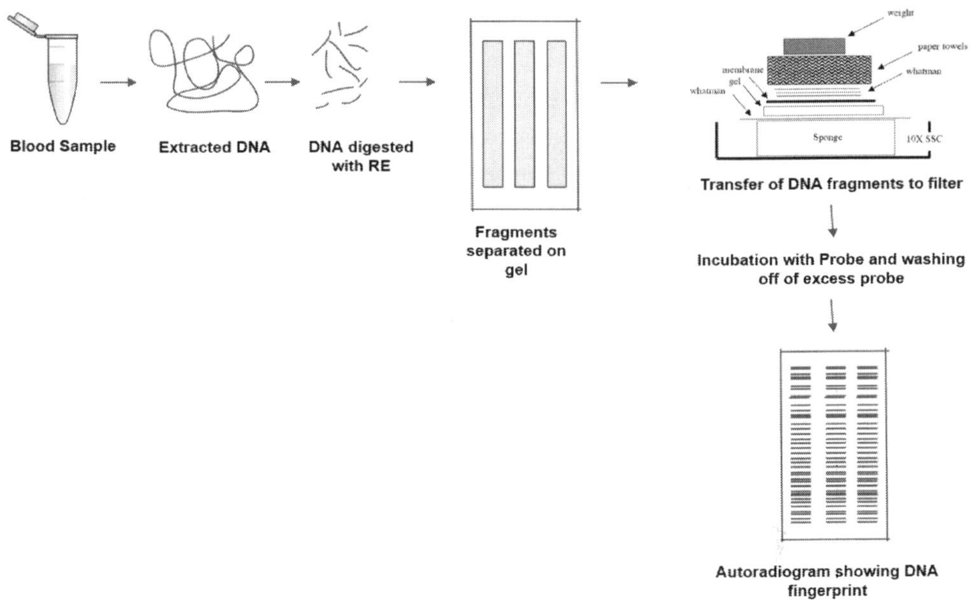

Figure 6e. The Procedure of DNA Fingerprinting.

The pattern thus obtained is called a DNA fingerprint. This process provides a highly unique pattern that has great discriminatory power but the procedure is very slow and cumbersome. With the invention of the Polymerase Chain reaction in 1983, the technique of DNA fingerprinting was modified to utilize the advantages of this "really superb technology" in Sir Alec Jefferys's own words. The Polymerase Chain reaction can be applied to this to make the technique automatic and quick. Moreover, it's very advantageous to use PCR to create a DNA fingerprint, as very little amount of DNA obtained from crime scenes which is otherwise insufficient for creating the DNA fingerprint using the classical method, is sufficient using this technique. But the problem was the use of minisatellites which are much bigger and hence the use of microsatellites began. These are smaller but have the same tandem repeat pattern so can be substituted for minisatellites.

The Procedure is modified when using PCR to create the DNA fingerprint. The Steps are

1. Extraction of DNA
2. Amplification of VNTRs using specific primers (Figure 6f)
3. Electrophoretic separation of amplicons to create the DNA fingerprint.

This technique which was invented by serendipity has become an extensively used technique in modern times. Sir Alec Jefferys was working on the myoglobin gene in whales when he accidentally obtained a pattern which was very complex in the first look, but was later understood by him.

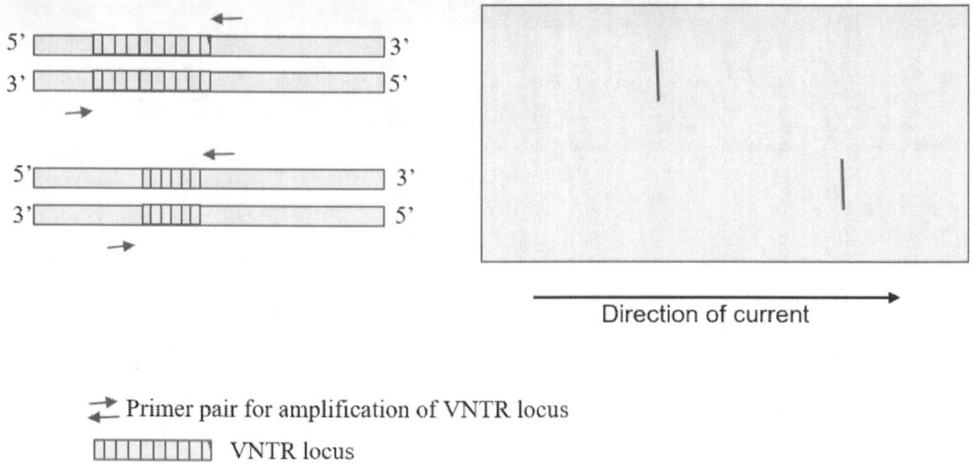

Figure 6f. PCR Primers for Amplification of VNTR Locus.

Dr. Lalji Singh is considered the "Father of Indian DNA Fingerprinting" as he is credited with the use of this technology in India. He worked towards establishing the Center for DNA Fingerprinting and Diagnostics (CDFD) in 1995. DNA Fingerprinting is used extensively in the US and UK identification of individuals. In the US a national DNA database - The **Co**mbined **D**NA **I**ndex **S**ystem (CODIS) has been created and is maintained by the FBI. It makes use of core loci for identifying individuals. It initially included 13 loci but 6 more were added later. Recently an MoU has been signed by CDFD and FBI and CODIS based DNA profile matching is now used by CDFD.

4. Application of DNA Fingerprinting

The earliest known use of DNA Fingerprinting was in March 1985 in solving a migration case. A Ghanaian lady's son on return to UK from Ghana was detained at the immigration questioning his identity as the son or a substitute relative. The fingerprint of the mother and three confirmed children were used to identify and prove that he was her son and not a substitute. In the same year this technique was used to solve a paternity case and to identify identical twins. The first ever use of DNA fingerprinting in criminal case was to solve a rape and murder case in 1986. DNA fingerprinting helped in successfully solving the case by identifying the real culprit and acquitting the suspect. This technique is now extensively used in USA and UK in identifying individual culprits based on their DNA fingerprint.

Since the DNA Fingerprint for each individual is unique, this technique is used to identify individual in

1. Paternity cases- Children share 50 per cent of their fingerprints with either parent. The DNA fingerprint of the mother, father, and child can be compared to prove paternity inclusion/exclusion. A Multilocus probe DNA Fingerprint of a child will have approximately half the bands matching

either parent. In contrast, in Single locus probe DNA fingerprint the child has one band matching the father and one matching the mother. In Figure 6g. (Panel A), it is clear that the child has few bands matching the father and few matching the mother. So it can be concluded that the child is the biological child of the couple.

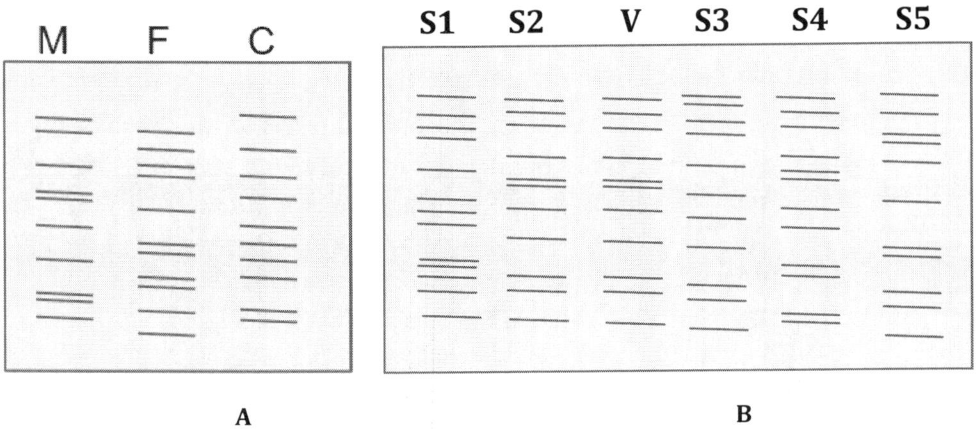

Figure 6g. Application of DNA Fingerprinting in Identification of Individuals. A is an example of paternity testing (M: mother, F: father and C: child) and B is identification of suspects.

2. Crime scenes- Victim and suspect identification can be done by collecting any available source of DNA from the crime scene and comparing the DNA fingerprint with people with well-defined motives in that particular case. In Figure 6g. (Panel B), the DNA fingerprint from the blood sample retrieved from a crime scene (V) was compared to 5 probable suspects(S1-5). The profile matches suspect 2, so in all probability, he is the culprit.

This technique can also be used in special cases of identification of missing persons, military cases, and dead bodies. The basis of all DNA fingerprinting based identifications can be summarized as

- ☆ Every individual inherits the satellite regions which are used to make the DNA fingerprint from their parents.
- ☆ The DNA fingerprint for every individual is unique except in the case of identical twins.

5. Advantages of DNA Fingerprinting

The technique of DNA Fingerprinting is very advantageous in the identification of individuals. Its advantages are

- It is a non-invasive method that is easy and has great discriminatory power.
- It is accurate, reliable, and affordable.
- It can make use of any nucleated cell available.
- DNA is more stable and not easily degraded by environmental factors, so the sample can be retrieved even after many days have elapsed.

6. Disadvantages of DNA Fingerprinting

Inspite of the above listed advantages, this technique has certain disadvantages too.

- The genetic variability due to varying allele frequencies of VNTRs in different ethnic groups and populations leads to problems.
- Loss of information due to extended time in sample retrieval
- The problem of chance matching in case of single locus probes needs to be ruled out by multiple single locus probes.
- The privacy, social and ethical issues related to this technique are a major concern

7. Questions

Multiple Choice Questions

1. The DNA fingerprint is identical for

a. Monozygotic twins b. Dizygotic twins

c. Siblings d. Cousins

2. The DNA fingerprint pattern of a child is

a. Exactly similar to that of both parents

b. 100 per cent similar to the mother's print

c. 100 per cent similar to the father's print

d. 50 per cent band similar to father and rest similar to mother

3. The technique of DNA fingerprinting was developed by

a. James Watson b. Hargobind Khurana

c. Kary Mullis d. Sir Alec Jefferys

4. The creation of a DNA fingerprint would normally employ all the techniques except

a. Southern Blotting b. Gel Electrophoresis

c. Western Blotting d. Restriction Digestion

5. A single locus probe DNA fingerprint has a maximum of

a. 1 band
b. 2 bands
c. Multiple Bands
d. None of the above

Short Answer Questions

1. **Elaborate on the principle of DNA fingerprinting.**
2. **Differentiate between single locus probe fingerprint and Multilocus probe fingerprint.**
3. **DNA Fingerprinting can be used to solve paternity claims. Justify with a hypothetical example.**
5. **Justify the statement that a maximum of two bands are obtained in Single locus probe DNA fingerprints.**

Long Answer Questions

1. **What is DNA fingerprinting? Explain the procedure and principle of this technique.**
2. **Explain the procedure of making a DNA fingerprint and discuss the advantages and disadvantages of this technique.**

REFERENCES

https: //le.ac.uk/dna-fingerprinting

https: //www.yourgenome.org/stories/the-eureka-moment-that-revolutionised-crime-solving

https: //thebiologist.rsb.org.uk/biologist-features/an-evening-with-alec-jeffreys-2

Gill P, Jeffreys AJ, Werrett DJ. Forensic application of DNA 'fingerprints'. Nature. 1985 Dec 12-18;318(6046): 577-9. doi: 10.1038/318577a0. PMID: 3840867.

Jeffreys AJ, Wilson V, Thein SL. Individual-specific 'fingerprints' of human DNA. Nature. 1985 Jul 4-10;316(6023): 76-9. doi: 10.1038/316076a0. PMID: 2989708.

Roewer, L. DNA fingerprinting in forensics: past, present, future. *Investig Genet* **4**, 22 (2013). https: //doi.org/10.1186/2041-2223-4-22

Saad R. Discovery, development, and current applications of DNA identity testing. *Proc (Bayl Univ Med Cent)*. 2005;18(2): 130-133. doi: 10.1080/08998280.2005.11928051

7

Study of Polymerase Chain Reaction

Dr. Madhu

ABSTRACT

Polymerase chain reaction is a laboratory technique, which is used to produce billions of copies of DNA sequences of interest using small quantity of DNA within few hours. This technique basically relies on short DNA sequences called primers to amplify the target region of the genome using heat stable polymerase under in vitro condition. These primers should anneal to the flanking sequences on both side of the target regions. This method involves cyclic increase and decrease of temperature to allow DNA polymerase to copy the target sequences. PCR has great medical and diagnostic applications. It is used for techniques like molecular cloning, site directed mutagenesis and many more. PCR has also contributed a lot to the study of metagenomics. Since the invention of PCR, different types of PCR have been developed for their application in different fields of science.

Keywords: *Polymerase chain reaction, dNTP, Primers, Template, Taq Polymerase.*

Learning Objectives

After the practical, students will be able to:

- ☆ Explain the process and concept of PCR
- ☆ Describe the process of Hybridization and Polymerization by Polymerase.

1. Introduction

Polymerase chain reaction is a technique that is used to amplify (produce millions of copies) of a selected region of DNA. It was invented by Kary B. Mullis in the year 1985. Any region of DNA can be taken provided border sequences of the selected stretch are known. Border sequences must be known because in order to carry out PCR two short oligonucleotides are used. These oligonucleotides act as primers to carry out DNA synthesis. They delimit the stretch of DNA which gets amplified. So, oligonucleotides must hybridize to these border sequences of DNA molecules, one to each strand of the double helix.

PCR is carried out in a thermal cycler, where reactants (DNA) are subjected to repeated cycles of heating and cooling to allow different temperature dependent reactions to go on. Amplification is carried out by using an enzyme called *Taq* polymerase (DNA polymerase 1 isolated from *Thermus aquaticus*). It is a thermostable enzyme.

To begin a PCR amplification, a reaction mixture containing template (DNA whose fragment needs to be amplified), primer (two short oligonucleotides), a mixture of free nucleotides and *Taq* polymerase is prepared. Reaction mixture is exposed to a high temperature to denature template DNA followed by cooling to allow the primers to bind the denatured DNA strands. *Taq* polymerase assembles new complementary strands using free nucleotides. Cycle of denaturation, hybridization and extension is repeated to produce multiple copies of DNA. As PCR progresses, the DNA generated during reaction is itself used as a template for replication, setting in motion a chain reaction in which the original DNA template is amplified exponentially.

2. Principle

The main significance of PCR is that it can amplify the selected stretch or fragment of DNA out of total DNA by using just a few reagents and multiple rounds of heating and cooling steps. During PCR, repeated rounds of replications take place. For replication to take place, double stranded template DNA is heat denatured to single stranded forms (in nuclear replication helicase converts double stranded DNA to single stranded form). As a result, when DNA is exposed to high temperatures, enzymes used here to carry out replication are required to be thermostable. *Taq* polymerase (DNA polymerase 1) is a thermostable enzyme isolated from *Thermus aquaticus*. This organism lives in hot springs, and many of its enzymes, including *Taq* polymerase, are thermostable. Thermostable means they are resistant to denaturation by heat treatment.

In addition to *Taq* DNA polymerase and template DNA, PCR requires unique short single stranded DNA called primers that bind, one to each strand of the double helix (Figure 7a). They are synthetic oligonucleotides. Primers delimit the size of DNA fragments to be amplified. Success of PCR depends upon the primer sequence. So, designing correct primers is crucial for this process.

Cycle 1

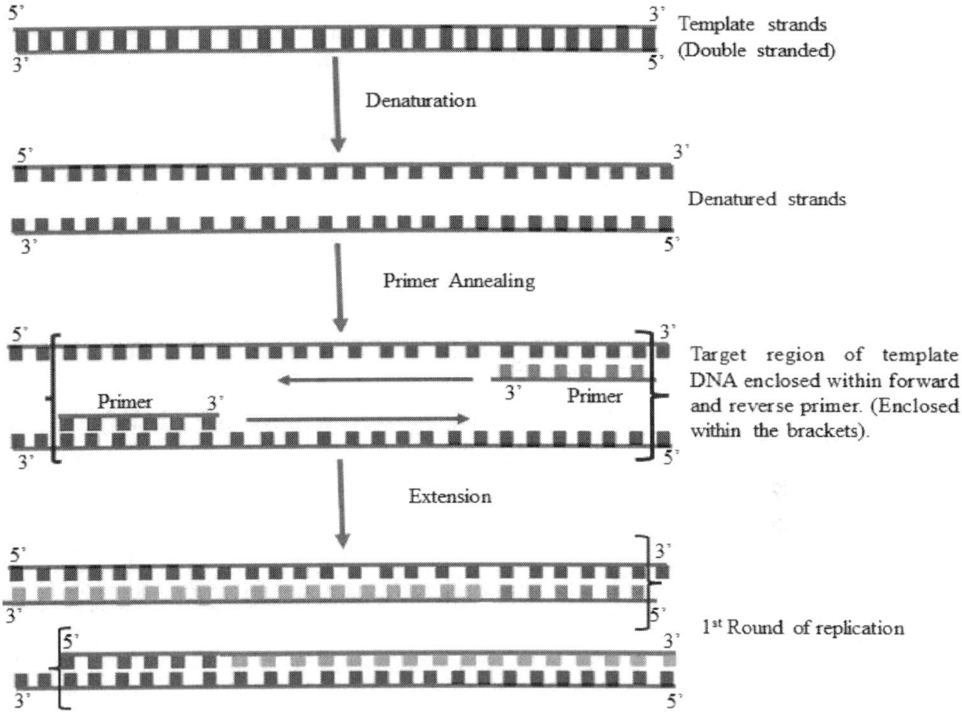

Cycle 2

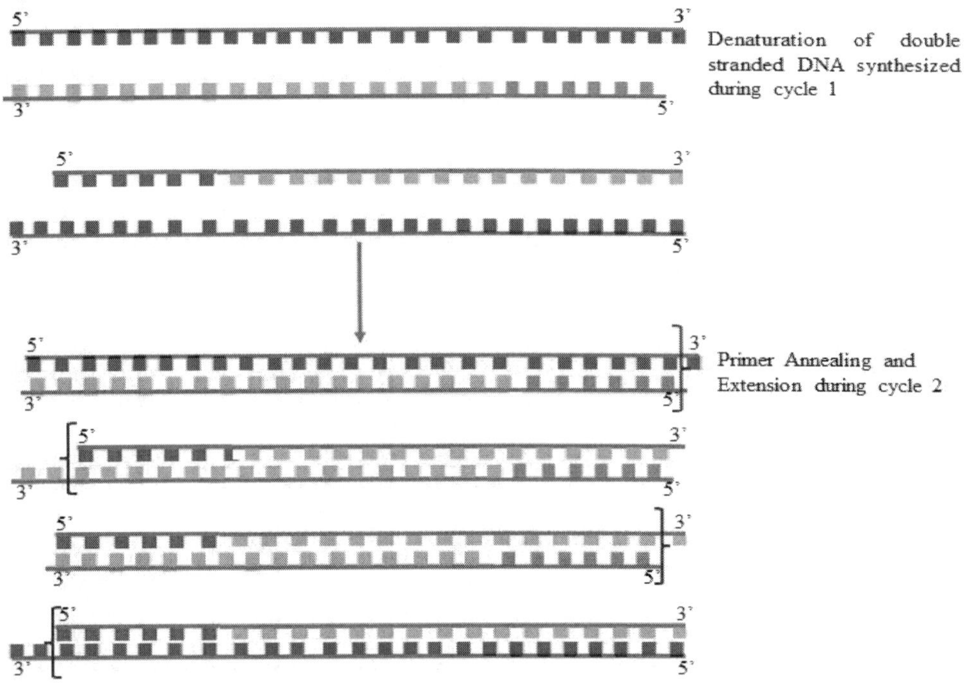

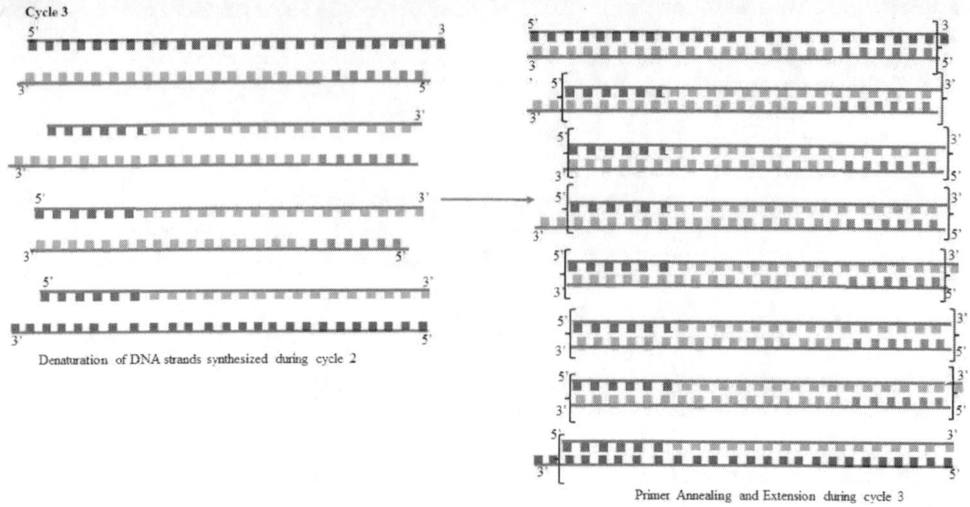

Figure 7a. Polymerase Chain Reaction.

Cycle 1: Doble stranded DNA is denatured and the two primers anneal to the denatured strands at the opposite ends of target region. Primers are extended by polymerase action to give DNA of variable length. Cycle 2: parent strand and newly synthesized strands from cycle 1 are denatured, which results in creation of 4 primers binding sites. Primers are extended by polymerase till the end of the template, synthesizing a precise copy of the target region. Cycle 3: DNA strands of cycle 2 are denatured and primer annealing and extension lead to production of DNA molecules that are identical to target region. With further progression of cycles, there would be exponential doubling of the target region while the original DNA strands and variably extended strands would become negligible in numbers.

There are few parameters which should be taken into consideration while designing primer sequences.

1. Length of 18-24 bases.
2. 40-60 per cent G/C content.
3. Start and end with 1-2 G/C pairs.
4. Melting temperature (T_m) of 50-60°C.
5. Primer pairs should have a T_m within 5°C of each other.
6. Primer pairs should not have complementary regions.

Free nucleotides are added to synthesize new complementary strands for the template strands by *Taq* polymerase. A suitable chemical environment is required by *Taq* polymerase to carry out PCR. This requirement is mediated by carrying a reaction in a suitable buffer. The buffer pH is usually between 8.0 and 9.5 and is often stabilized by Tris.HCl. For *Taq* DNA polymerase, a common component in the buffer is potassium ion (K^+) from KCl, which promotes primer annealing. $MgCl_2$ is added in a buffer to stabilize DNA duplex structure.

There is a limitation to the size of DNA which can be amplified efficiently by PCR. The length to be amplified should not be greater than 3 kb. Standard PCR technique can amplify upto 10 kb. Even fragments upto 40 kb can be amplified but require special methods.

There are three basic steps of PCR:

1. **Denaturation**- To amplify DNA, like in any other replication, the two strands of the template DNA are separated from each other (Fig7a). This occurs by heat denaturation. The dsDNA templates are heated to a point (around 94 °C) where the hydrogen bonds break between the base pairs. This results in the separation of the two DNA strands.
2. **Annealing**- The separated strands are going to act as template for the synthesis of complementary strands. Before replication gets started it is essential for primers to hybridize with the template DNA at the correct site. This is achieved by bringing down the temperature from denaturation temperature. This process of hybridization of primer with the template is called Annealing. The temperature is dropped to a range in which the forward and reverse primers bind to the template. The annealing temperature should be low enough to allow hybridization between primers and template but should be high enough to prevent mismatch binding. Temperature should be 1-2 °C less than melting temperature of Primer. $T_m = (4\times[G+C]) + (2\times[A+T])$.
3. **Extension**- Once the primer has bound to the template, *Taq* polymerase can initiate the assembly of free nucleotides into the complementary strands by its 5'-3' polymerase activity. This process is called extension. Polymerase extends the primer to form nascent DNA at temperatures ranging from 68-75 °C. This temperature range is considered to be the optimum temperature for maximal activity of *Taq* polymerase. Extension temperature is also determined based on the AT content of the sequence to be amplified. *Taq* polymerase polymerizes DNA approximately at the rate of 1Kb per minute.

These processes are repeated multiple times (typically 25-35 cycles), and because each newly synthesized strand serves as a template for the primers in the next cycle, there is an exponential doubling in the copy number of the region of interest.

3. Requirements

Thermal cycler (Figure 7b), microfuge tube (vol. 200 µl), Pipette, Nuclease free water, Template DNA, Forward and Reverse primers, deoxynucleoside triphosphates (dNTPs) mix, *Taq* polymerase, Enzyme buffer.

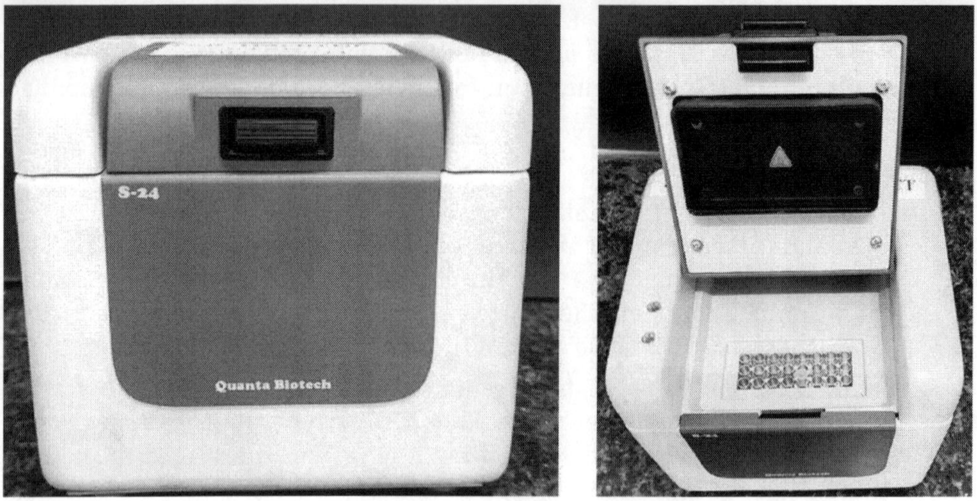

Figure 7b. A Thermal Cycler to Carry Out PCR.

4. Procedure

4.1. Pre-experiment Steps (Necessary steps needed to take before preparation of reaction mixture)

I. Reaction mixture is prepared in an eppendorf tube also called PCR tube having 200 μl volume. It should be sterilized (nuclease free).
2. All the components of the mixture should be ice thawed well before use.
3. All the components should be stored in ice only during the use and once done they should be stored in a deep fridge.
4. Preparation of reaction mixture should be done in ice only.
5. To make the reaction mixture, all the components should be added in the order as mentioned in Table 2.

Table 2: List of Components and their Concentrations to be Added to the Reaction Mixture

Name of the Component	*Final Concentration in the Reaction Mixture*
Nuclease free water	To adjust the final volume (as per calculations)
Taq Buffer	1×
dNTP mix	200 μM
Forward and reverse primer	0.4 μM (each)
Template DNA	50-100 ng
Taq Polymerase	0.06 U/μl

6. Required volume of micropipette should be ready.
7. Volume of nuclease free water is calculated before starting the preparation of reaction mixture by using the formula: (Total reaction Volume-sum of volumes of rest of the components).

4.2. Preparation of Reaction Mixture

1. Two microfuge tubes are taken. One will be having a PCR reaction mix with template and other one will be a negative control for the reaction mix without the template DNA.
2. To the microfuge tube, calculated amount of nuclease free water is added followed by addition of *Taq* buffer, dNTP mix, forward and reverse primer and then the template DNA at given concentration.
3. Once all these components have been added, *Taq* polymerase is added as the last component of the reaction mixture.
4. Once added, the reaction mixture is vortexed gently to mix all the components with the enzyme.

4.3. PCR set-up

1. Thereafter, microfuge tube containing PCR mixtures are placed in a programmed thermal cycler (Figure 7b)
2. Cycling conditions can vary depending upon the fragments to be amplified. The given cycling condition shown in Table 3 can be used for amplification of 1.5 kb of 16S *rRNA* gene of bacteria using their universal forward and reverse primers.
3. Once PCR amplification gets completed, the product can be resolved and visualized on agarose gel containing ethidium bromide by gel electrophoresis.

   (*Note*: Please refer agarose gel preparation and electrophoresis (sections 4.2 and 4.3) in 'Isolation of genomic DNA' experiment)

Table 3. Cycling Condition for Amplification of 16S *rRNA* Gene (Conditions can vary depending upon amplicon and primers chosen)

*Initial Denaturation Temperature	94 °C for 5 min	Only at the beginning
Denaturation Temperature	94 °C for 1 min	30 cycles of Denaturation, annealing and extension
Annealing Temperature	55 °C for 30 Sec	
Extension Temperature	72 °C for 2 min	
**Final Extension Temperature	7 min at 72 °C	Only at the end

Note: *Purpose is to ensure that all complex double stranded DNAs (template DNA) are converted into single strands for amplification. ** Purpose is to ensure polymerization of full length amplicon and a good yield of the target DNA. It also ensures filling up of incomplete ends of polymerized strands. It also helps in TA cloning.

5. Result and Discussion

On agarose gel an amplicon of X Kb would be visualized (depending upon the template and primers chosen). Under ideal condition, a clear sharp band corresponding to the amplicon size should be obtained (shown in Figure 7c) with no other bands. A clear sharp band with correct size of amplicon indicates the specificity of both the primers to the template.

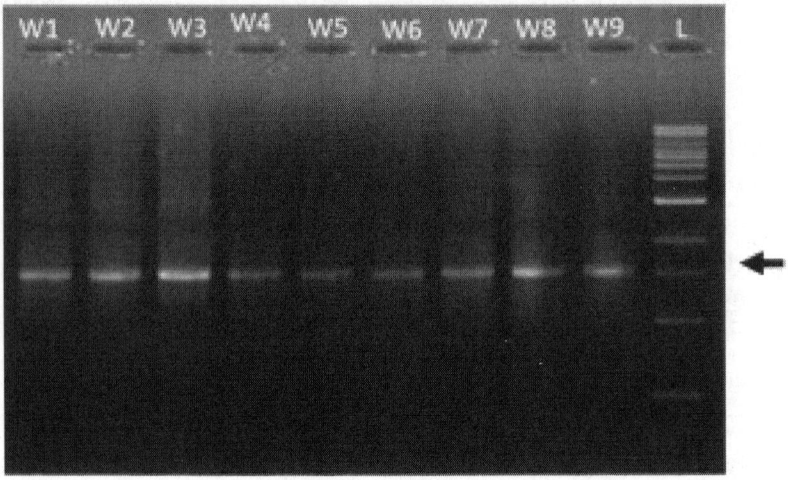

Figure 7c. A 0.8 per cent Agarose Gel Showing Amplification of 16S *rRNA* Gene of Bacteria. Lane L: 500 bp ladder, W1-W9: Amplicon of 16S *rRNA* gene using their universal primers.

Sometimes, some additional bands get visualized whose size is even less than 100 bp. These bands are referred to as Primer Dimer (PD). PD consists of two primer molecules that have hybridized with each other because of strings of complementary bases in the primers. In addition to Primer Dimer, sometimes few additional bands may appear other than the main amplicon. These bands correspond to the non-specific binding of primers to the template, which gives non-specific amplification and product of varied sizes.

6. Precautions

- All the microtips and microfuge tubes being used for PCR should be nuclease free.
- Enzymes should always be kept on ice.
- All the components should be added at the optimum concentration
- Once PCR gets completed, PCR products should be stored at 4°C until they are used.
- Agarose gel containing EtBr should be handled always with hand wearing gloves.

- ✰ PCR products should always be loaded on gel containing ladders in the adjacent well.
- ✰ MQ water should also be sterilized before use.

7. Questions

Multiple Choice Questions

1. Who invented PCR?

a. Kary Mullis b. Brigitte Boisselier

c. Jacob Monod d. Hargobind Khurana

2. Which of the following enzyme is needed to carry out PCR?

a. Alkaline phosphatase b. Kinase

c. *Taq* polymerase d. Telomerase

3. In which cycle of PCR amplicons corresponding to the size of target fragment start appearing?

a. 1st Cycle b. 2nd Cycle

c. 3rd Cycle d. 4th Cycle

4. What is the optimum length of primer?

a. 10-20 b b. 18-24 b

c. 30-45 b d. 40-55 b

5. ***Taq* polymerase is isolated from which bacteria?**

a. *E.coli* b. *Streptococcus*

c. *Thermus aquaticus* d. *Rhodococcus*

Short Answer Question

1. Define the following

- ✰ **Polymerase chain reaction**
- ✰ **Primer**
- ✰ **Denaturation**
- ✰ **Annealing**
- ✰ **Extension**

2. Why denaturation, annealing and extension are carried out at different temperatures?

3. Why do we use *taq* polymerase for polymerase chain reaction to be carried out?

4. **What are the parameters that need to be taken into consideration while designing a primer?**
5. **Which technique is used for visualization of PCR amplicon?**

Long Answer Question

1. **Discuss the different phases of polymerase chain reaction with their significance.**
2. **Discuss the role of all ingredients being added to the PCR mix.**
3. **What are the possible sources of error if we are not getting a PCR amplicon of correct size or not getting any amplicon? Discuss them.**

REFERENCES

https: //en.wikipedia.org/wiki/Polymerase_chain_reaction

https: //www.addgene.org/protocols/primer-design/

https: //en.wikipedia.org/wiki/Primer_dimer

https: //www.thermofisher.com/in/en/home/life-science/cloning/cloning-learning-center/invitrogen-school-of-molecular-biology/pcr-education/pcr-reagents-enzymes/pcr-cycling-considerations.html

Gene Cloning and Analysis, T A Brown, 7th Edition

8

Western Blotting

Dr. Sunita K. Yadav

Dr. Shailesh Kumar R. Yadav

ABSTRACT

Western blotting is a common technique used in cell and molecular biology. It is also called protein immunoblotting because an antibody is employed to specifically detect its protein/antigen. Individual proteins can be identified from a complicated mixture of proteins isolated from cells by utilizing a Western blot technique. To accomplish this job, the approach employs three major steps: (1) Separation of proteins based on size, (2) transfer of proteins to a solid matrix, and (3) visualization of the target protein using the appropriate primary and secondary antibodies. This chapter will aim to explain the theory and process of Western blotting along with some troubleshooting tips.

***Keywords**: Nitrocellulose Membrane, Blotting, Proteins, Enzyme-conjugated secondary antibody, Electrophoresis.*

Learning Objectives

After the experiment, student will be able to:

- ☆ Understand the concept of Western blotting
- ☆ Perform the experiment efficiently
- ☆ Realize the importance of the Western blotting technique

1. Introduction

Western blotting is an analytical technique used in cell and molecular biology to detect specific proteins from a mixture of proteins. The technique involves separation of proteins based on their molecular weight and then transference of proteins on blotting matrix (nitrocellulose or Polyvinylidene difluoride membrane). The tagged antibodies (conjugated with enzyme or fluorescent or radioactive probe) are then used to identify protein of interest on membrane. The quantity of protein decides the thickness of the band *i.e.*, thickness of the band is directly proportional to the amount of protein. A standard protein marker can be run alongside the gel to estimate the concentration of protein. The explanation of various steps of Western blotting is given below (Figuer 8a):

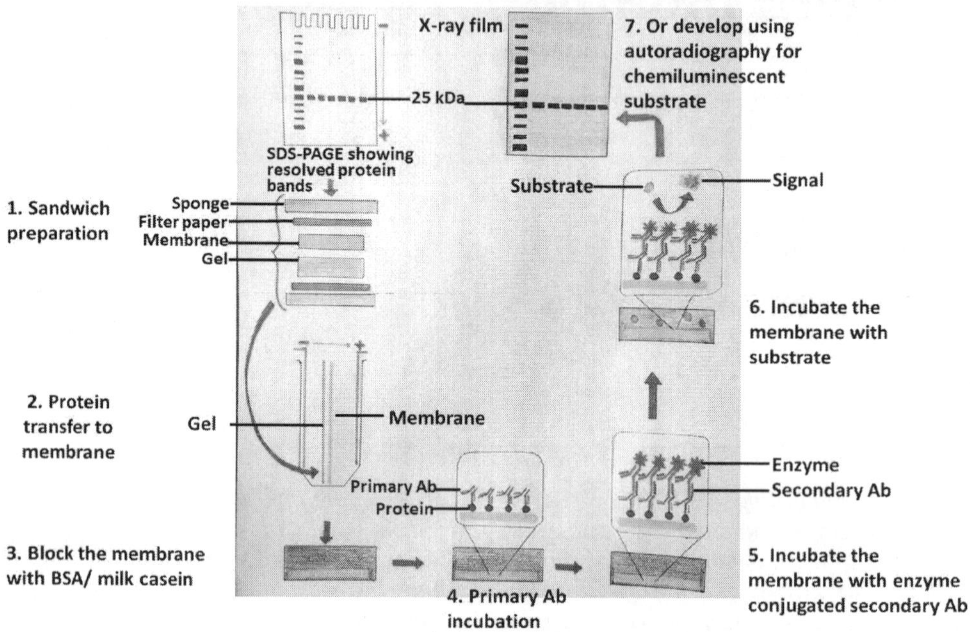

Figure 8a. Steps of Western Blotting Technique (Illustrator: Aakriti Singh, Department of Botany, University of Delhi).

Sodium Dodecyl Sulphate - Polyacrylamide Gel Electrophoresis (SDS-PAGE)

It is a standard technique to separate the proteins on the basis of their molecular weight. The polyacrylamide gel is the result of polymerization of acrylamide and bis-acrylamide (N,N'-Methylenebisacrylamide). The nucleation to polymerize is initiated by addition of ammonium persulphate and TEMED (N,N,N′,N′-tetramethylethylenediamine). The pores of the gel help in separation of proteins. The pore size depends on the amount of acrylamide and crosslinker used. The higher percentage gels have smaller pore size while lower percentage gels have larger pore size. The percentage of gel is crucial factor that decides rate of migration and degree of separation between proteins. A positive control and protein marker should be loaded along with the samples while running the gel. The positive control eliminates any confusion in interpreting the result in case of no bands formed in the sample lane. It assures that the correct protocol is followed. The protein marker helps in determining the size of the protein present in the sample. The protein loading dye is always used to load the samples in the wells. The dye helps in monitoring the progress of the gel.

Transfer of Proteins on Membrane

The proteins are transferred on membrane from the gel in the presence of electric field. The membrane should be directly over the gel for protein transfer without even any air bubble between them. The gel and membrane, both are sandwiched between sponge pads and Whatman paper that act as adsorbent material. The entire sandwich arrangement is immersed in transfer buffer for electrical supply. For the transfer, it is necessary that gel should face the negative electrode (cathode) while membrane should be towards positive electrode (anode). The negatively charged protein are repelled by cathode and attracted to the membrane which is positively charged, towards anode side. Several types of membrane are available for protein transfer. The most popular are nitrocellulose and polyvinylidene difluoride (PVDF) membranes.

Development of Western Blot

After the proteins are transferred to the membrane, several steps are carried out to develop the protein bands. To remove the background and prevent non-specific binding of antibodies to protein, blocking buffer (skimmed milk or BSA) is used. The primary antibody specific to protein are added at a specific dilution otherwise high concentration of antibody can lead to development of non-specific bands. The dilution can be determined beforehand by ELISA or some other technique. The incubation time of antibody depends on the affinity of antibody to antigen/protein and its dilution. A longer incubation is required for low affinity and diluted antibodies. During incubation period, continuous gentle agitation of antibody is advised for homogenous exposure to proteins/antigens and to prevent uneven binding. The washing step is imperative after every incubation to prevent interference of reagent with each other and to remove non-specific binding of antibodies. The secondary antibody acts as detection antibody generally and their dilution is prescribed by the manufacturer. The secondary antibodies are conjugated with enzyme (generally alkaline phosphatase or horseradish peroxidase) and addition of chromogenic substrate results in conversion of it into colored products that is precipitated at the position of protein band (s). The developed bands are interpreted and analyzed.

2. Principle

The Western blotting technique is based on antigen-antibody interaction. The term 'blotting' refers to the transfer of molecules from gel to the membrane and subsequent processing and detection are carried out on the membrane. Western blotting was introduced by Towbin et al.,1979. Western blotting is also called immunoblotting because an antibody is used for detecting proteins. The specificity of antibody to the antigen/protein helps in detection of a specific protein in the mixture of proteins.

The technique uses SDS-polyacrylamide gel electrophoresis (SDS-PAGE) to resolve the mixture of proteins in the sample (Native gel separates proteins based on their conformation and structure while denaturing gel separates on the basis of molecular weight of polypeptide). The separated proteins on gel are transferred on to membrane (nitrocellulose or PVDF membrane) and allowed to bind to specific antibody. The antibody is conjugated with enzyme or fluorescent or radioactive probe and gives color or blot on the membrane or on the X-ray film of specific band of protein. The intensity and size of observed band give an idea about the amount of a specific protein. The Western blot procedure is a sensitive method that can detect as low as 1 ng of protein.

3. Requirements

- ☆ Apparatus: SDS-PAGE assembly, blot Transfer assembly (Figures 8b, 8c and 8d).
- ☆ Glasswares and plastic wares: Measuring cylinder, beaker, conical flask, *etc.*
- ☆ Others: Nitrocellulose membrane, Antibodies (Primary antibody, secondary or detection antibody), Western blue substrate
- ☆ Chemicals and buffers: Acrylamide, N,N'-Methylenebisacrylamide, TEMED, Ammonium persulphate, Sample buffer, running buffer, electrode transfer buffer, 1×PBS, 1×PBST, 3 per cent BSA (Blocking Buffer)

4. Procedure

The procedure of Western blotting is divided in to three major parts and a detailed explanation is given below.

4.1. Separation of Proteins on SDS-PAGE (Resolution of proteins)

4.1.1. Preparation of SDS-PAGE

Polyacrylamide gels are produced by polymerization of two-components, acrylamide and bis-acrylamide. Bis-acrylamide is a cross-linking agent for the gels. The polymerization of the acrylamide is initiated by the addition of ammonium persulphate (0.1 per cent) along with TEMED. The molecule separation in gel

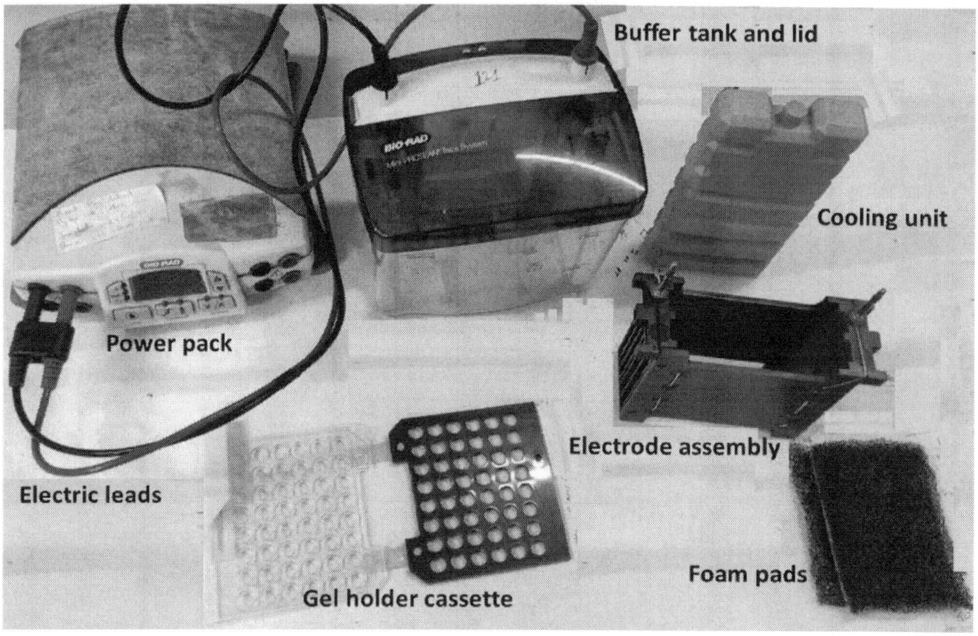

Figure 8b. Western Blot Equipment.

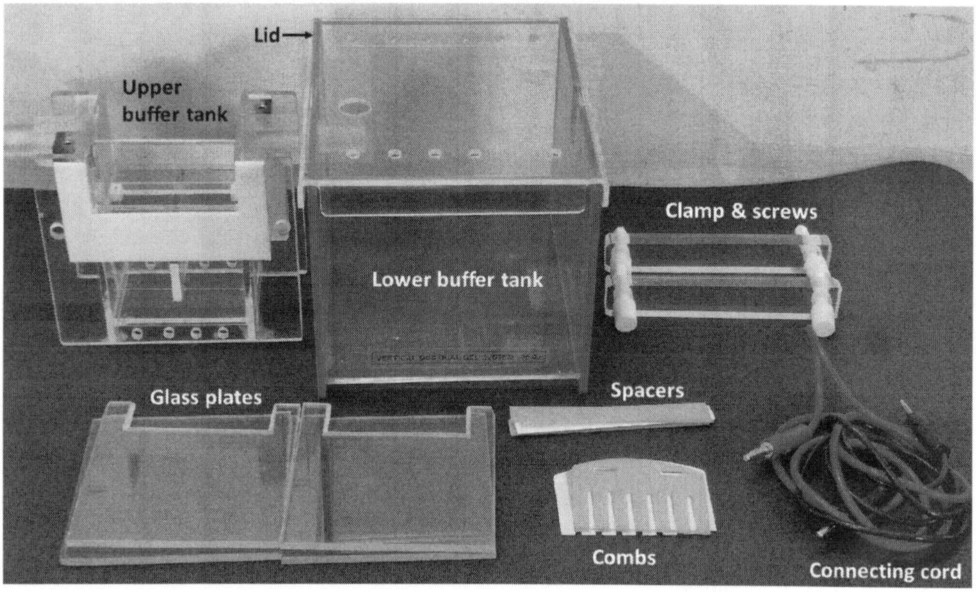

Figure 8c. SDS-PAGE Equipment Showing all the Parts (Pic courtesy: Dr. Nitin Bhardwaj).

depends on the pore size of the gel formed which depends on the proportion of acrylamide and bis-acrylamide used to form the gel.

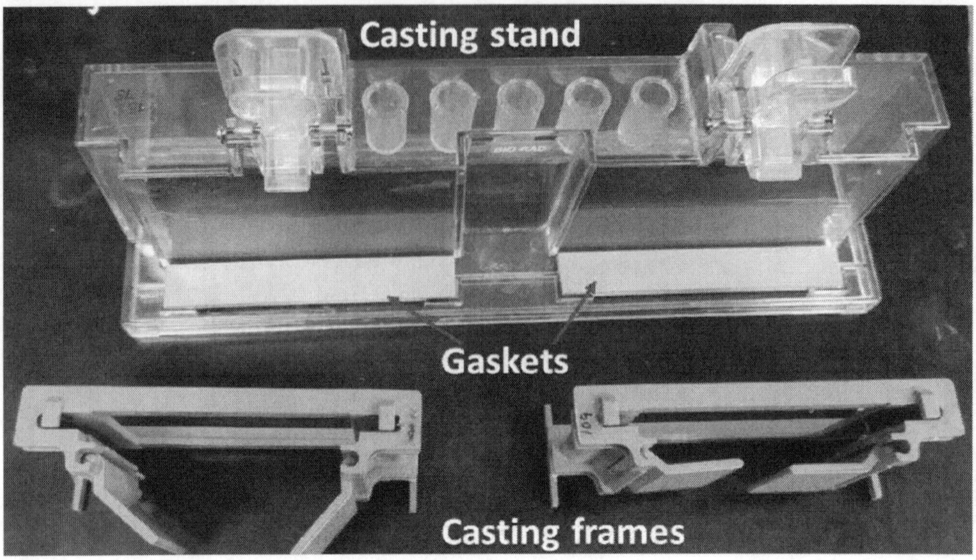

Figure 8d. Gel Casting Assembly.

4.1.2. Resolution of Proteins on SDS-PAGE

1. Wipe the clean glass plates with alcohol and assemble them with the spacers.
2. Make SDS-PAGE gel by pouring resolving and stacking gel. The percentage of gel depends on the size of proteins, that needs to be separated.

   (Note: APS and TEMED were added to initiate polymerisation, just before pouring the gel)
3. For resolving gel, mix appropriate amounts of stock solutions in the proportions given in Table 4.
4. Pour the resolving gel slowly into the assembled plates avoiding formation of air bubbles and layered with water (or isopropanol) without disturbing the poured solution. After 15–20 min when the solution polymerizes, remove water from the top and overlay the stacking gel solution on it and put the combs to make grooves for well. The composition of stacking gel is given in Table 5.
5. Leave the gel for complete polymerization for at least 15 min.

Table 4. Composition of Resolving Gel

Components	*Percentage of Separating Gel*	
	10 per cent	*12 per cent*
Acrylamide stock (30 per cent)	3.3 ml	4 ml
1.5 mM Tris.HCl, pH 8.8	2.5 ml	2.5 ml
10 per cent SDS	0.1 ml	0.1 ml
DD Water	4 ml	3.3 ml
10 per cent APS	0.1 ml	0.1 ml
TEMED	0.004 ml	0.004 ml
Total volume	10 ml	10 ml

6. Sample preparation: Mix protein sample (1–100 μg) with sample buffer (final concentration 1× = 0.0625 M Tris.HCl, pH 6.8, 2 per cent SDS, 10 per cent glycerol, 5 per cent β-mercaptoethanol (for reducing) and boil for 10 min in a boiling water bath.
7. Load the samples carefully in the wells of prepared polyacrylamide gel and electrophoresed at 20 mA (constant current) in running buffer (0.025 M Tris.HCl, pH 8.3, 0.192 M glycine and 0.1 per cent SDS).
8. Run a protein marker to detect the size of the protein.

Table 5. Composition of Stacking Gel

Components	*Volume*
Acrylamide stock (30 per cent)	0.83 ml
1 M Tris.HCl, pH 6.8	0.63 ml
10 per cent SDS	50 μl
Water	3.9 ml
10 per cent APS	50 μl
TEMED	5 μl
Total volume	5 ml

4.2. Transfer of Proteins to Membrane

Transfer the protein samples (bacterial cell lysates) resolved on 12 per cent SDS-PAGE onto the nitrocellulose membrane using electrode transfer buffer [25 mM Tris.HCl, pH 8.3, 192 mM glycine, 20 per cent (v/v) methanol] using transfer blot apparatus. The Sandwich arrangement of gel in Western blotting apparatus is shown in Figure 8e.

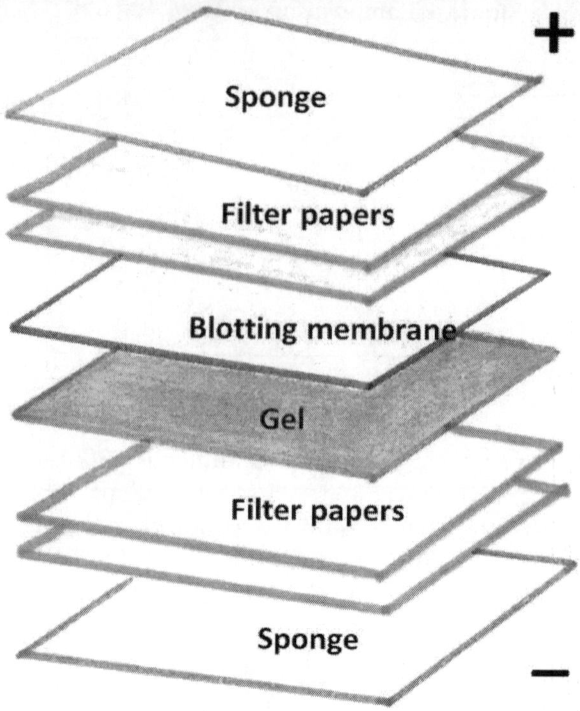

Figure 8e. The Sandwich Arrangement of Gel in Western Blotting (Illustrator: Aakriti Singh, Department of Botany, University of Delhi).

4.3. Binding of Antibody to Membrane

1. Non-specific sites were blocked by incubation in 3 per cent BSA (or 5 per cent skimmed milk) in 1×PBST (0.15 M PBS, pH 7.3, and 0.05 per cent Tween 20) at 4 °C O/N followed by 2 washes with 1×PBST.
2. The membrane was incubated with the primary antibody (anti-His-tag monoclonal antibody or anti-protein antibody raised in animal such as mice at the dilution indicated in the legend) for 1 h at 37°C followed by three washes with 1×PBST (10 min each).
3. The membrane was then incubated with enzyme conjugated (horseradish peroxidase or alkaline phosphatase) secondary antibody [(raised in different species against Fc domain of primary antibody (for example, Rabbit anti-mouse secondary antibody)] at 1:10,000 dilution (or as mentioned by the manufacturer) for 1 h at 37°C, followed by three 1×PBST washes.
4. The immunoreactive bands were visualised by the addition of appropriate substrate solution according to enzyme conjugated to secondary antibody

[5-bromo-4-chloro-3-indolyl-phosphate (BCIP) and nitro blue tetrazolium (NBT) for alkaline phosphatase or TMB (3, 3', 5, 5'-tetramethylbenzidine) or 3,3' diaminobenzidine (DAB) substrate for HRP enzyme]

5. Stop the reaction by removing the substrate solution and subsequently washing it with distilled water.

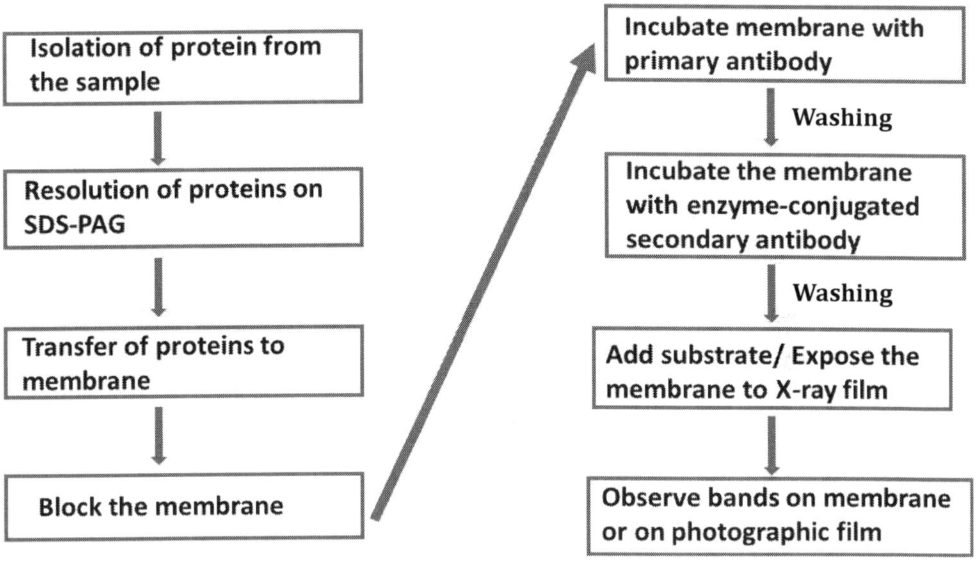

A Flowchart Showing Steps of Western Blotting.

5. Result

The proteins are observed as dark bands on the membrane. The size of protein can be estimated by the protein ladder. The intensity of the band is directly proportional to the high amount of protein (Figure 8f).

6. Discussion

Western blot is an important technique used for protein detection and quantification. Although, the Western blotting procedure is simple, few problems in protein band detection arise sometimes that lead to unpredicted results. The membrane used for protein detection sometimes show variations in results such as unusual bands, no bands or very faint bands. The faint, patchy or blurry bands can be resulted from high voltage application or a bubble formation while transferring the bands from gel to membrane. Therefore, it is desirable to run the gel at low voltage and in cool environment and taking proper care (no bubble formation between gel and membrane) while transferring bands to membrane. Also, fresh running buffer should be used as sometimes it can also lead to faint bands.

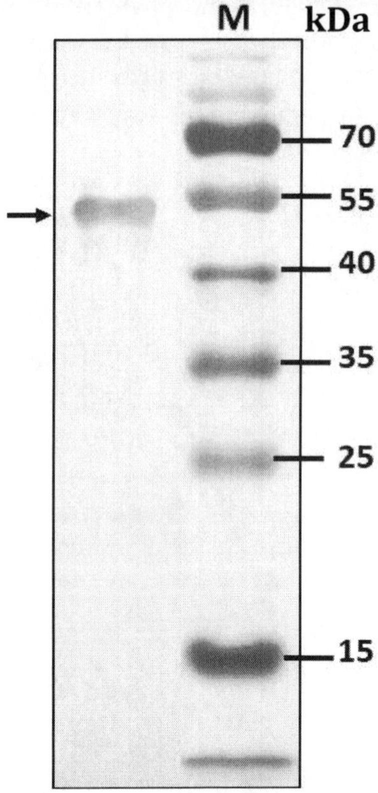

Figure 8f. A Western Blot Membrane Showing an Immunoreactive Band (Marked by arrow) between 40 kDa and 55 kDa Bands of Protein Ladder (pre-stained).

Sometimes unusual bands are seen at a different location. These bands can be formed by protease degradation leading to reducing the size of the protein and hence location on Western blot. The proteases are generally present in old samples therefore it is advisable to use only fresh samples that are protected on ice before loading. The high position of protein is observed in those cases where the protein fails to be denatured and still is in its quaternary structure. Therefore, it is advisable to denature the protein in lysis buffer in presence of β-mercaptoethanol followed by boiling for some time (~10 min.).

The high background on the membrane can be due to high concentration of antibody or non-specific binding of antibody to the membrane. Hence, the membrane should be blocked with 3-5 per cent BSA before incubating with antibody. It will result in blocking non-specific sites of membrane that will then not be available for antibody binding. Sometimes, high antigen/antibody concentration, old buffers, less washing time after incubations of antibody or increased exposure time led to high background. Therefore, it is important to standardize the conditions for optimum protein/antigen, washing time and antibody dilution before proceeding with experiment. Also, exposure time can be observed intermittently to reduce the background.

As PVDF membrane is hydrophobic in nature so transfer buffers will not be able to penetrate the membrane homogeneously. Therefore, it is advised to activate the PVDF membrane with absolute methanol for 30 sec and then in transfer buffer for 1 -2 min (until it sinks). otherwise there will be problem in transfer of proteins and binding on membrane. Methanol activates the membrane and helps in hydration with transfer buffer and increases binding capacity for proteins. It has also been reported that PVDF membranes give high background as compared to nitrocellulose membrane, so membrane selection can be done according to the expected result.

It is advised to use shaker at each step of Western blotting as it will help homogenous covering of membrane by the reagents including BSA and antibodies.

7. Applications

- It helps to identify a specific protein from the mixture of proteins on the basis of size and interaction with specific antibody.
- The level of expression of a particular protein can also be estimated based on the intensity of band.
- The expression of target protein can be determined and compared in various tissues.
- The response of a particular protein after treatment with a drug or during diseased condition can be evaluated.
- It is also used for diagnostic purposes in a wide range of applications such as detection of HIV infection, mad cow disease, Hepatitis B infection *etc.*

8. Precautions

- Cover your face with a mask and wear gloves while preparing acrylamide gel.
- There should not be any bubble formation between gel and membrane while transferring proteins from gel to membrane.
- The pH of buffers and all the reagents should be properly checked.
- The dilution of antibodies and concentration of protein should be optimized to give better results.
- Blocking reagents (such as BSA or skimmed milk) should be used to decrease the background.
- PVDF membrane should be activated with absolute methanol (1-2 min.) and then kept in transfer buffer (until it sinks) prior to use.
- Take care of all the precautions and steps that are discussed in the discussion part.

9. Questions

Multiple Choice Questions

1. **Which of the following act as an initiator for acrylamide polymerization?**
   a. APS
   b. TEMED
   c. Both
   d. None
2. **Which of the following is neurotoxic?**
   a. APS
   b. TEMED
   c. Acrylamide
   d. Polyacrylamide
3. **BSA and casein are used during Western blotting to**
   a. Block the non-specific sites on membrane
   b. Block the unwanted primary antibodies
   c. Block the unwanted proteins
   d. All of them
4. **The principle of Western blotting includes**
   a. Separation of proteins according to molecular weight
   b. Blotting of proteins to membrane
   c. Antigen-antibody interactions
   d. All of them
5. **Primary antibodies are**
   a. Specific to antigen
   b. Raised in the same species as antigen
   c. Bound with enzyme
   d. All of them
6. **The arrangement of different things in Western blotting from the positive anode is**
   a. Sponge-filter paper-membrane-gel-filter paper-sponge
   b. Filter paper-sponge-membrane-gel-sponge-filter paper
   c. Sponge-filter paper-gel-membrane-filter paper-sponge
   d. Filter paper-sponge-gel-membrane-sponge-filter paper

Short Answer Questions

1. **What is the principle of Western blotting?**
2. **Why blocking is required before adding primary antibody?**
3. **Why detection/secondary antibody should be linked with enzyme?**

4. **Define following terms**
   - ☆ **Blot**
   - ☆ **Blocking**
5. **Give the sequence in which membrane, gel and Whatman paper are arranged in the blotting apparatus.**
6. **What is the reason for a high background on developed membrane?**
7. **Why is it important to use protein marker in the SDS-PAGE gel?**
8. **If the size of protein is 55 kDa, what should be the per cent of SDS-PAG?**
9. **What could be reasons of no band on the developed membrane?**
10. **What could be the reasons for a faint band on the developed membrane?**
11. **Why do stacking and separating gels have different compositions?**

Long Answer Questions

1. **Give steps of Western blotting.**
2. **Explain the advantages of Western blotting.**
3. **Differentiate between Western blotting, Southern blotting and Northern blotting.**

REFERENCES

Becker JM, Caldwell GA, Zachgo EA (2007) Biotechnology- A laboratory course. Academic Press. An imprint of Elsevier, San Diego, California.

https: //www.sigmaaldrich.com/IN/en/technical-documents/protocol/protein-biology/western-blotting/western-blotting

https: //www.thermofisher.com/in/en/home/life-science/protein-biology/protein-biology-learning-center/protein-biology-resource-library/pierce-protein-methods/overview-western-blotting.html

Maniatis T, Fritsch EF, Sambrook J (1982) Molecular cloning: A laboratory manual. Cold Spring Harbor Laboratory, Cold Spring Harbor, NY

Towbin H, Staehelin T, Gordon J. 1979. Electrophoretic transfer of proteins from polyacrylamide gels to nitrocellulose sheets: procedures and some applications. *Proc Nat Acad Sci USA* **76**: 4350– 4354.

Mahmood T, Yang PC. Western blot: technique, theory, and trouble shooting. *N Am J Med Sci*. 2012;4(9): 429-434. doi: 10.4103/1947-2714.100998

9

To Demonstrate the Technique of Southern Hybridization

Dr. Sunita K. Yadav

ABSTRACT

The first blotting technique was developed by biologist E.M. Southern in 1975 to identify specific DNA sequences in the DNA fragment. In his honour, the technique was named Southern blotting. The technique is based on the detection of DNA fragments on the membrane using specific probe. The technique is used in molecular biology, immunogenetics and other DNA analysis method to identify/detect the gene/DNA of interest from the mixture of DNA samples.

***Keywords**: Electrophoresis, Probe, DNA fragment, Hybridization, Membrane.*

Learning Objectives

After the exercise, students will be able to:

- ☆ Differentiate different types of blotting
- ☆ Understand the principle of Southern blotting
- ☆ Apply the knowledge of Southern blotting in various applications

1. Introduction

The Southern blot is named after Edwin Southern, who invented the first method for analysing DNA in 1975. To assess genetic diversity, commonly referred to as polymorphism, this approach makes advantage of numerous fundamental features of DNA. The procedure starts with the cutting of genomic DNA with restriction endonucleases into smaller fragments followed by electrophoresis to separate the fragments based on their size. DNA fragments are then transferred to a blotting paper and incubated with specific probes to identify/select the desired target fragments. It's most commonly used to diagnose viral and bacterial illnesses. Southern blotting is a technique to identify a specific DNA in rDNA technology. This feature is utilised to diagnose neonatal sickness and genetic disorder. It is also valuable in the investigation of mutation and gene rearrangement. Phylogenetic studies, paternity and maternity analyses, forensic studies, and personal identification all use DNA identification because of its precision.

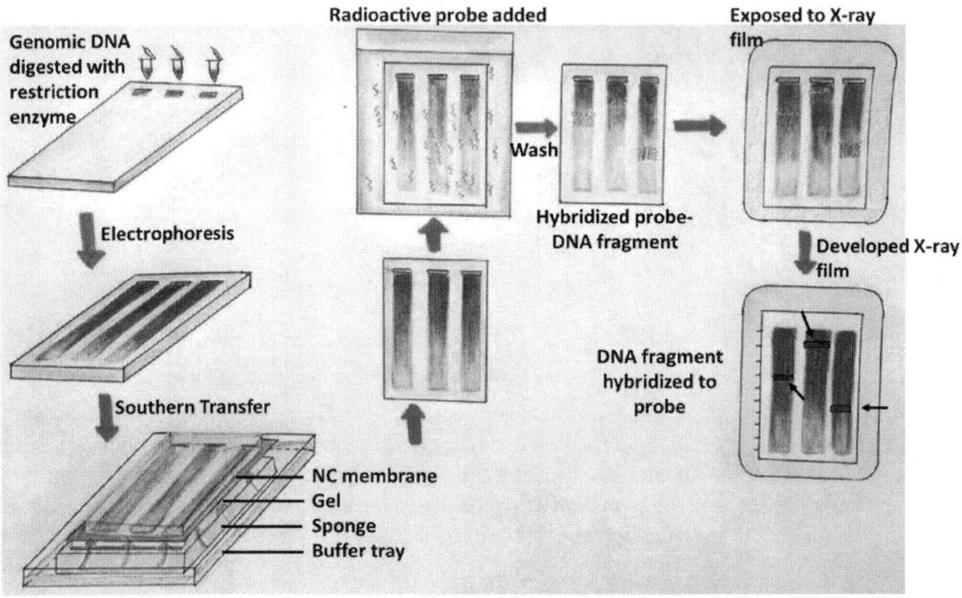

Figure 9a. Steps of Southern Blotting (Illustrator: Aakriti Singh, Department of Botany, University of Delhi) (Modified from Hood, 2006).

2. Principle

The technique is based on electrophoresis-blotting-hybridization methods. In the process, the isolated/amplified DNA fragments separated by gel electrophoresis are transferred to a membrane (nitrocellulose/nylon). The DNA fragments on the membrane are then detected by hybridization using a probe. Hybridization results in formation of double-stranded DNA molecules from a single-stranded target DNA and single-stranded DNA probe as both are complementary to each other. The binding results in some signal in the form of a band that can be easily detected by various methods (Figure 9a).

3. Procedure

1. **Isolation of DNA:** The DNA is isolated from the target cells/culture of animal cell lines/culture of bacteria by following standard protocol of genomic DNA extraction and then purified if need arises.
2. **Restriction enzyme digestion or DNA fragmentation:** The high molecular size DNA is cut in to smaller fragments using restriction endonucleases. Depending on the size requisite for experiment, one or more restriction enzymes can be employed to attain required fragments.
3. **Size separation by gel electrophoresis:** The digested DNA fragments in the previous step are separated by agarose gel electrophoresis. The negatively charged DNA molecules move towards positively charged electrode of apparatus assembly based on their size. The DNA ladder loaded estimates the size of the fragments.
4. **Depurination and denaturation:** First, partial depurination of DNA fragments is done using diluted HCl to promote higher efficiency of transfer to membrane in the subsequent steps. Depurination is followed by denaturation wherein a mild alkali such as 10 N solution of NaOH is used to denature the DNA (separation of double stranded DNA in to single-stranded DNA) to allow it for hybridization with probe in next steps. Neutralization to prevent re-hybridization of single strands of denatured DNA is done by using NaCl.
5. **Blotting:** The denatured fragments of DNA from gel are transferred onto membrane (Nitrocellulose/Nylon) by placing it on the gel which is placed on buffered saturated filter paper. Dry filter papers in the quantity of 3-4 are then placed on top of the membrane. The weights (0.5-1 kg kg) can also be put on the filter paper. The DNA fragments are pulled towards the membrane by capillary action of buffer resulting in replica formation of gel bands on membrane.

   Note: Please refer Figure 8e in 'Western blotting' technique for the arrangement pattern of gel and membrane during blotting step)
6. **Baking:** After 15 h, then nitrocellulose membrane is removed from the blotting apparatus and subjected to baking at 80 °C for 2 h/120 °C for 30 min or exposed to UV light to fix the DNA onto the membrane.
7. **Hybridization:** The ssDNA on the membrane is then exposed to a ssDNA fragment acting as hybridization probe. The hybridization probe has a specific sequence by which the target DNA needs to be identified. The probe DNA is labelled with radioactivity or chemiluminescence or chromogenic dye to detect the target DNA. The probes are highly specific

and can selectively bind with a resolution of 1 in a million to the intended target fragments.

8. **Washing step:** The membrane after hybridization, is thoroughly washed with a buffer to remove the unbound probe or non-specifically bound probe.
9. **Detection of hybridized probes:** The hybridized probes are detected by exposing the membrane to a photographic film in case of a radioactive or fluorescent probe or by the development of color in case a chromogenic probe is used.

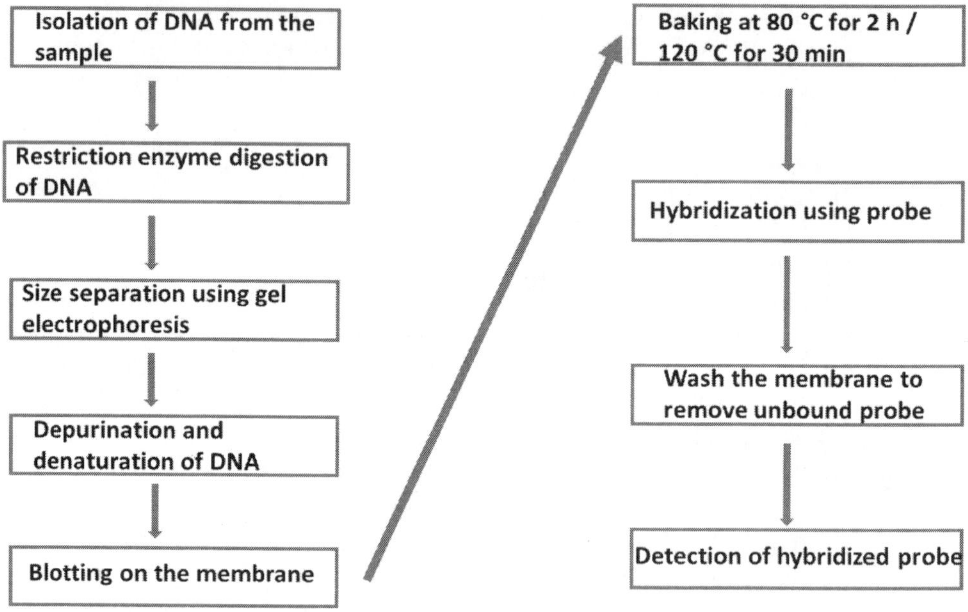

Steps of Southern Blotting.

4. Result

The probe is able to bind to a target sequence and a high intensity band is visible on the film after developing it with autoradiography. The DNA marker has resolved on the membrane and a comparison of DNA marker with the band shows that the size of bands are around 14 kb, 18 kb and 40 kb in length (Figure 9b).

5. Applications

- ☆ Identification of specific DNA from a mixture of DNA
- ☆ Constructing Restriction Fragment Length Polymorphism (RFLP) maps.
- ☆ Detection of mutations, deletions or gene rearrangements in DNA

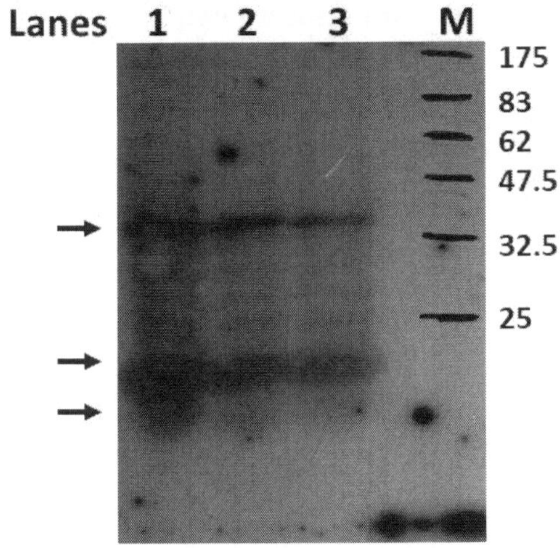

Figure 9b. A Southern Blot Showing Bands at different Sizes (Pic courtesy: Prof. Aparna Dixit's laboratory, JNU, New Delhi).

- ☆ Used in forensic studies to identify the criminal or for paternity and maternity test using DNA fingerprinting (VNTR)
- ☆ Identification and detection of a foreign gene in transgenic individual
- ☆ Mapping of restriction enzyme sites.
- ☆ In the diagnosis of infectious diseases (Virus/bacteria).
- ☆ Prenatal diagnosis of genetic disorder and prognosis of cancer.
- ☆ In phylogenetic diseases.

6. Questions

Multiple Choice Questions

1. Which is the first step of southern technique among following steps?

a. Electrophoresis b. Blotting

c. Restriction digestion d. Hybridization

2. The probes that can be used for southern technique is/are

a. RNA b. DNA

c. Protein d. RNA and DNA

3. Radioactive probes have generally:

a. Labeled N b. Labeled O

c. Labeled P d. Labeled H

4. The cross linking of DNA-membrane can be done by:

a. UV treatment b. Autoclaving

c. Baking at a higher temperature d. All of them

5. Which one is not an application of southern blotting:

a. Identification of DNA b. RFLP recognition

c. Mapping d. Antibody screening

Short Answer Questions

1. What is the principle of Southern hybridization?

2. Define the following terms:

- **Probe**
- **Hybridization**

3. What is the sequential arrangement of blotting stuffs in Southern hybridization?

4. Give characteristics of a good probe.

Long Answer Questions

1. Write applications of Southern blotting?

2. How Southern blotting is different from Western blotting?

REFERENCES

Becker JM, Caldwell GA, Zachgo EA (2007) Biotechnology- A laboratory course. Academic Press. An imprint of Elsevier, San Diego, California.

https: //www.sigmaaldrich.com/IN/en/technical-documents/protocol/protein-biology/gel-electrophoresis/southern-and-northern-blotting

https: //www.thermofisher.com/in/en/home/life-science/dna-rna-purification-analysis/nucleic-acid-gel-electrophoresis/southern-blotting.html

Ian A. Hood (2006),The mycology of the Basidiomycetes

Maniatis T, Fritsch EF, Sambrook J (1982) Molecular cloning: A laboratory manual. Cold Spring Harbor Laboratory, Cold Spring Harbor, NY

Primrose S B, R M Twyman and R W Old. Principles of Gene Manipulation. 6th edn.

Southern E. Southern blotting. Nat Protoc. 2006;1(2): 518-25. doi: 10.1038/nprot.2006.73. PMID: 17406277.

10

Northern Blotting

Dr. Sunita K. Yadav

ABSTRACT

Northern blotting technique is used to study gene expression by detecting the corresponding mRNA on membrane. Northern blotting technique was developed by James Alwine, David Kemp and George Stark in 1977 at Stanford University. The name of technique is based on its similarity of the process with Southern blotting. The identification, quantity, activity, and size of a specific gene can all be determined using this method. This blotting technique can be used to assess the growth of tissue or organism. The abundance of an RNA fluctuates during distinct stages of differentiation and morphogenesis, which can be detected using this method. At the molecular level, it also aids in the detection of abnormal, diseased, or infected conditions.

Keywords: *RNA, Hybridization, Electrophoresis, Probe.*

Learning Objectives

After the experiment, students will be able to:

- ☆ Perform Northern blotting efficiently
- ☆ Understand the principle of Northern blotting
- ☆ Design a specific probe to screen the RNA sequence of interest
- ☆ Differentiate among different types of blotting

1. Introduction

James Alwine, David Kemp, and George Stark of Stanford University invented the Northern blot technique in 1977. The name "Northern blotting" comes from its resemblance to the initial blotting technique, the Southern blot, which was named after biologist Edwin Southern. The main distinction is that the Northern blot examines RNA rather than DNA. The Northern blot, also known as an RNA blot, is a molecular biology technique that detects RNA (or extracted mRNA) in a sample to evaluate gene expression. By establishing the specific gene expression rates throughout differentiation and morphogenesis, as well as in aberrant or pathological settings, northern blotting can be used to examine cellular control over structure and function. Northern blotting is a technique that uses electrophoresis to size-separate RNA samples before detecting them using a hybridization probe that matches part of or the complete target sequence. The phrase 'Northern blot' refers to the transfer of RNA to the membrane by capillary action from the electrophoresis gel in its strictest sense. Northern blotting is the term used to describe the entire procedure. Northern blotting allows researchers to compare the expression profiles of specific genes across pathogen infections, levels of environmental stress, developmental stages, tissues, and organs, as well as over the course of treatment. When compared to "normal" tissues, the Northern blotting approach can be used for detecting upregulation of oncogenes and downregulation of tumor-suppressor genes in cancer.

2. Principle

The Northern blotting technique starts with the separation of RNA samples based on their sizes using electrophoresis method. The electrophoresis separates RNA based on the charge present on the molecules that are directly proportional to the size of RNA. The separated RNA is blotted on the membrane because gels are fragile and further steps cannot be proceeded using gel. The blotted RNA on membrane is subjected to hybridization using radioactive or labelled probe or oligonucleotides to detect the complementary RNA sequences. The hybridization results in a signal which can be detected by developing the membrane or using chemicals (Figure 10a).

3. Procedure

3.1. Gel Preparation for RNA Resolution

RNA samples are separated on agarose gel containing formaldehyde to denature the RNA. Small RNA or micro-RNA sequences can be resolved on polyacrylamide gel and urea can be used as a denaturing agent. Ethidium bromide is used to confirm the proper resolution of RNA on gel while RNA ladder and ribosomal units are used for size estimation.

3.2. Designing the Probe

The probe used for hybridization of RNA is complementary to the whole

RNA or a small region of interest. An RNA, DNA or oligonucleotides (~ 25 bp) complementary to target RNA is used as a probe. *In vitro* produced RNA probes are preferred over *in vivo* produced probes as latter can be denatured during the process of Northern blotting. The cDNA probes can be labelled with radioactive substances or with enzymes (alkaline phosphatase or horseradish peroxidase) in chemiluminescence case.

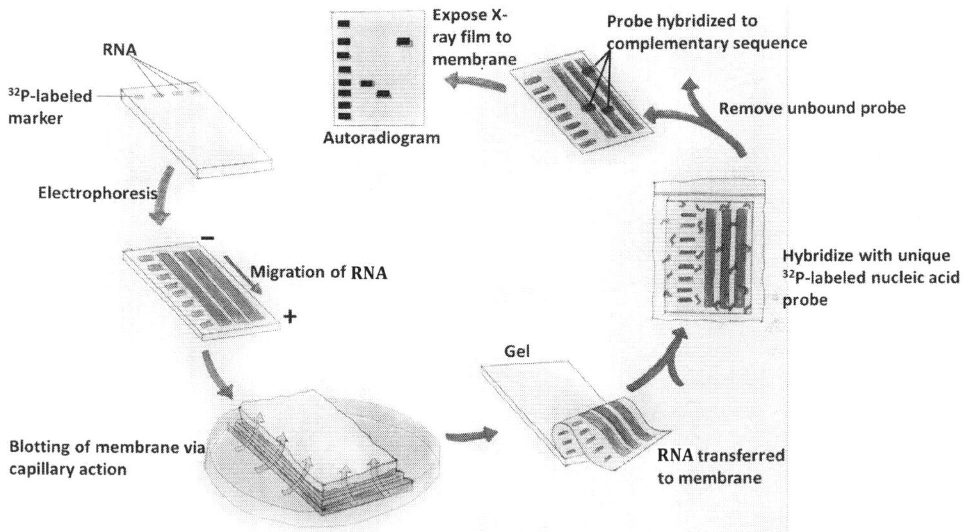

Figure 10a. Steps of Northern Blotting
(Illustrator: Aakriti Singh, Department of Botany, University of Delhi).

3.3. Northern Blotting Steps

1. The tissue/culture sample is homogenized to release RNA from the cell and RNA can be isolated by using T-chain of polynucleotides as mRNA has poly-A tail.
2. The RNA sequences in the samples are separated by electrophoresis on agarose gel.
3. The separated RNA sequence is transferred to membrane (nylon/ nitrocellulose) by capillary action. The RNA stick to membrane due to ionic interaction.
4. The transfer process requires the stack formation of agarose starting from the bottom followed by blotting membrane and a stack of paper towels pressed by weight (generally a glass plate) The entire formulation is kept in solution of running buffer.

   (*Note*: Please refer to Figure 8e in 'Western blotting' technique for the arrangement pattern of gel and membrane during blotting step)

5. The transferred RNA on the membrane is fixed by UV radiation.
6. The probe (specific for gene of interest) is then allowed to hybridize with RNA on membrane. The annealing temperature can be reduced by using formamide.
7. The unwanted probe is removed by washing step.
8. The blot membrane is washed to remove unwanted probe and the bound labelled probe is detected by autoradiography or chemiluminescence. An X-ray film gives dark bands as a result.

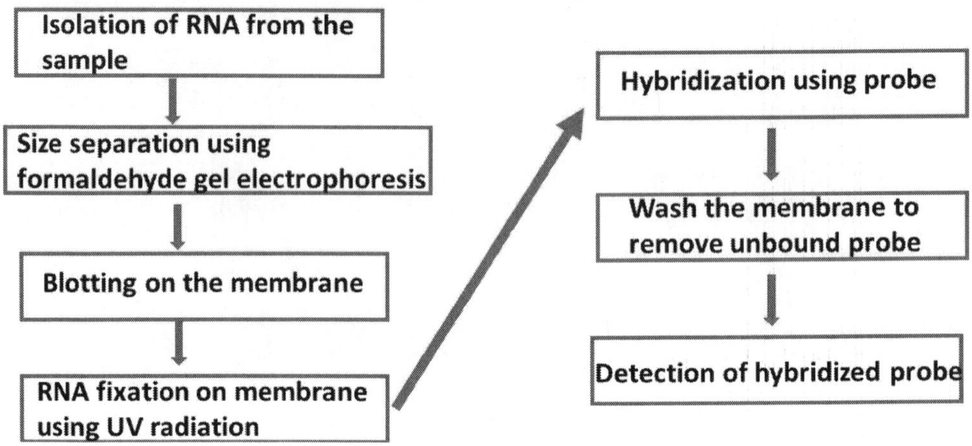

Steps of Northern Blotting.

5. Applications

- ☆ It is used to study gene expression.
- ☆ It reveals the identity, quantity, activity, and size of the particular gene.
- ☆ It can also be used for the assessment of gene expression during the growth of a tissue or organism.
- ☆ The RNA changes in different stages of differentiation and morphogenesis of an organism can be identified.
- ☆ It helps in the identification of mutated, diseased or infected conditions at the molecular level by analyzing the changes in mRNA expression.

6. Questions

Multiple Choice Questions

1. **The Northern blotting is used to assess:**
   a) Gene expression during morphogenesis
   b) Gene expression during differentiation

   c) Expression of gene during different stages of life
   d) All of the above

2. **The isolation of mRNA is based on:**
   a) Poly A tail   b) Poly T tail
   c) Uracil as base   d) All of the above

3. **The annealing temperature of RNA-DNA can be lowered by using:**
   a) SSC solution   b) Transfer buffer
   c) Formamide   d) NaCl

4. **Which types of probes are used for Northern blotting:**
   a) Short RNA   b) Short DNA
   c) Short oligonucleotides   d) All of the above

5. **The restriction enzymes are not used in:**
   a) DNA fingerprinting   b) Cloning
   c) Restriction mapping   d) Northern blotting

Short Answer Questions

1. **How RNA is fixed on the membrane after transfer?**
2. **Give characteristics of a good probe?**
3. **How RNA gel is made?**

Long Answer Questions

1. **What is the principle of Northern blotting?**
2. **Differentiate between Northern blotting and Southern blotting.**
3. **Describe applications of Northern blotting**

REFERENCES

https: //www.genome.gov/genetics-glossary/Northern-Blot

https: //www.sigmaaldrich.com/IN/en/technical-documents/protocol/protein-biology/gel-electrophoresis/southern-and-northern-blotting

Maniatis T, Fritsch EF, Sambrook J (1982) Molecular cloning: A laboratory manual. Cold Spring Harbor Laboratory, Cold Spring Harbor, NY

Primrose S B, R M Twyman and R W Old. Principles of Gene Manipulation. 6th edn.

Part B

Immunology

1

Demonstration of Lymphoid Organs in Rat

Dr. Sushma Bhardwaj

ABSTRACT

The innate immune cells provide the first line of defense against pathogens. Antigen presenting cells present the antigens to the lymphoid cells which co-ordinate the adaptive response, generate the memory cells and provide immunity against future infections. Components of the immune system are organized in lymphoid organs which are the integrated part of various organs and are distributed throughout the body. The primary and the secondary lymphoid organs coordinate the differentiation, proliferation, and maturation of the lymphocytes and provide the site of interaction between the antigen and the antibodies. Bone marrow, thymus, spleen, lymph nodes are the major lymphoid organs. These organs can be displayed in a dissected rat for studying the location and anatomy of the lymphoid organs. The focus of this chapter is to study the anatomy and the histological organization of the major lymphoid organs.

Keywords: *Lymphoid organs, Lymph node, Spleen, Thymus.*

Learning Objectives

After performing this experiment, students will get an overview of the lymphoid organs and the associated terms. They will be able to:

- ☆ Differentiate between the functions of primary and secondary lymphoid organs.
- ☆ Identify and locate the major lymphatic organs in dissected rat.
- ☆ Differentiate between the various lymphoid organs based on their structure, histology and functions.

1. Introduction

Our innate immune system provides the first line of defense against pathogen entering our biological system. A successful immune response is the result of the timely coordination between the specialized cells and the organs located in different part of our body forming the lymphoreticular system. The lymphoreticular system comprises of cells of diverse morphology which are distributed in different organs and tissues of the body. The two major components of the lymphoreticular system are:

Lymphoid Cells

Lymphoid cells include lymphocytes and plasma cells and are responsible for providing specific immunity.

Reticuloendothelial Cells

The reticuloendothelial cells form a universal immunologic cell system present in the vertebrates. Reticuloendothelial system (RES) is a heterogenous group of phagocytic cells and plasma cells present in reticular connective tissues, primary monocytes and macrophages. They trap and clears the soluble substances and particles such as bacteria, toxins, exogenous antigens, and immune complexes, circulating in the body. It forms an important part of the immune system. Due to their phagocytic role, the RES is also referred as mononuclear phagocytic system. These cells are responsible for non-specific immunity. They kill the pathogenic invaders and remove them from the body.

The Lymphoid Organs (Figure 1a)

The specialized organs or tissues which provide the appropriate microenvironment for specialization, maturation, or specific functions of the immune cells, are known as ***lymphoid organs.*** The lymphoid organs can be functionally classified in two groups:

I. Primary (Central) Lymphoid Organs

They provide the site for maturation of the lymphocytes. Thymus and bone marrow are the primary lymphoid organs in mammals. The hematopoietic stem cells from the fetal tissues or the adult bone marrow are lodged in the primary lymphoid organs, where they differentiate and mature into lymphocytes. The lymphocytes are the precursors of the immunologic function and are also the regulators and effectors of immunity. Most lymphocytes are short-lived with 5 – 7 days of shelf life while some T cells and B cells have a longer life span ranging from months to years. The primary lymphoid organs produce a reservoir of reactive lymphocytes with specific antigen receptors. These are also the site for the elimination of the cells with receptors for self-antigens to prevent autoimmunity.

II. *Secondary (Peripheral) Lymphoid Organs*

Secondary lymphoid organs provide the site for interaction of the mature lymphocytes with the antigens that are trapped from various tissues or vascular spaces. Lymph nodes, spleen, mucosa- associated lymphoid tissue (MALT) are the secondary lymphoid organs. Secondary lymphoid organs are the site of cytokine-cell and other cellular interactions. They coordinate the interaction of antigen and antigen specific lymphocytes resulting in production of effector and memory cells. The vascular system provides the interface and the micro-environment for this interaction.

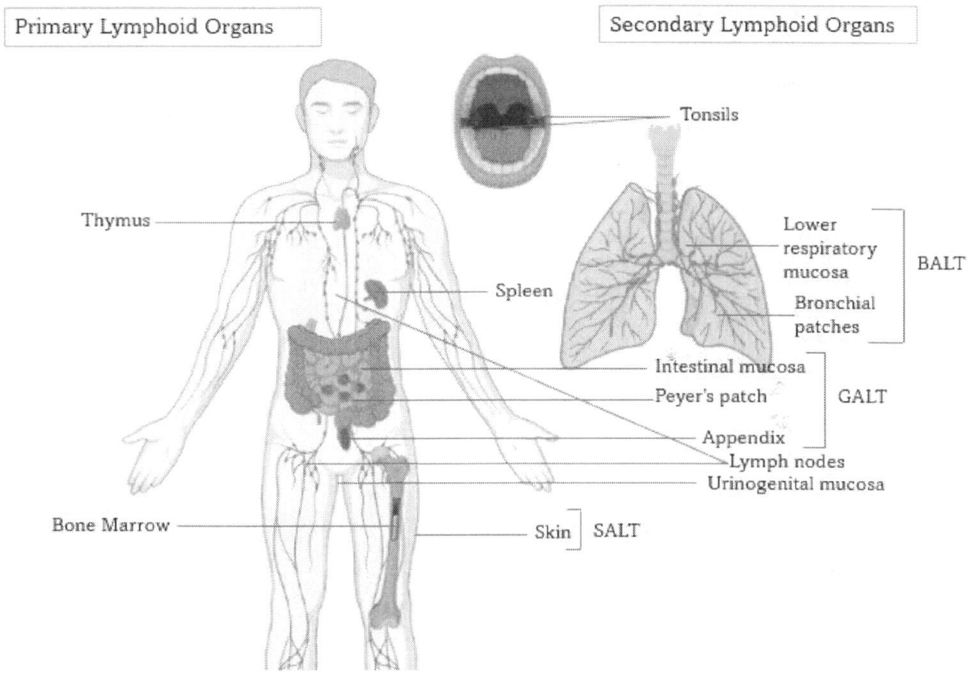

Figure 1a. Primary and Secondary Lymphoid Organs Present in Man.

Development of Lymphocytes

The lymphoid cell line originates from the fetal pleuripotent hematopoietic stem cells found in the yolk sac, liver, and the bone marrow. In the embryo, hematopoiesis begins in the yolk sac but as the embryogenesis proceeds, the function is taken over by spleen, lymph nodes, and the liver. In the later stages of gestation, the bone marrow of the fetus takes over the hematopoietic function and become responsible for the differentiation of the blood cells and the lymphatic cells. This function is retained by the adult bone marrow which produces all the blood cells including the RBCs, WBCs, and the platelets via myeloid lineage. Whereas, the lymphoid progenitor cells of the hematopoietic stem cells differentiate into the T cells, B Cells and the natural killer (NK) cells. The microenvironment around the stem cells in the lymphoid organs lead to their differentiation to lymphoid pathway.

The immune cells that mature in the thymus are called **T lymphocyte (T cells),** and those produced and maturate in the bone marrow are the **B lymphocyte (B cells)**. The sites where the plasma cells produce antibodies in the lymph nodes are called **germinal centers or follicles**. The small B lymphocytes are present in the primary follicles. After the antigenic stimulation, secondary follicles are formed containing many large dividing lymphocytes and plasma cells.

The lymphocyte proliferation and maturation in the primary lymphatic organs is independent of the exogenous antigen exposure. Extra-thymic maturation of T-lymphocytes also takes place in the small intestine, but it is not known if the process is induced by exogenous antigens. On the other hand, the proliferation and activation of the lymphocytes in the secondary lymphoid organs, *i.e.*, in spleen, lymph nodes, MALT is antigen dependent. Antigen-dependent lymphocyte proliferation also occurs in thymic medulla and in bone marrow.

2. Theory

Rat is the model organism to study the anatomy and histological organization of mammalian tissues.

There are four main species of rats that are commonly used in India for dissection:

Rattus rattus, R. bengalensis, R. norvegicus are the wild type rats found in the cities and rural areas. *R. rattus and R. begalensis* are the large sized rats while *R. norvegicus* is medium sized. *Rattus norvegicus albinus* (albino rats) are the white laboratory rats. Narcotized rats are dissected for the anatomical and histological studies of mammals.

In laboratory, lymphoid organs of mammals are studied in the rats. The position and anatomy and histological structure of mammalian lymphoid organs are similar. Rats knocked out with chloroform are dissected to expose the lymphoid organs. Table 6 shows the various lymphoid tissues and their functions.

3. Materials Required

- ☆ Rat anesthetized by chloroform
- ☆ Dissecting tray
- ☆ Scissors
- ☆ Scalpel
- ☆ Needle

Table 6: Lymphoid Tissues and their Functions

Sl.No.	*Lymphoid Organ and Compartments*	*Types of Cells Present*	*Functions*
1	**Bone marrow**	Hematopoietic cells organized as islands in the adipose tissues, plasma cells and mature leukocytes	Differentiation of hematopoietic stem cells, antibody synthesis and storage of memory cells
2.	**Thymus**		
	Cortex	Fine reticular epithelial cells, macrophages, mature T-cells	Generation of T-cell competence, T-cell receptor rearrangement, positive and negative selection of T-cell, removal of auto-reactive cells
	Medulla	Thick reticular epithelial cells, macrophages, dendritic cells, T-lymphocytes	T-cell competence (negative selection), thymic hormone secretion, antigen presentation
	Cortico-medullary zone	Immature and mature lymphocytes	Entrance of bone-marrow derived cells and exit of mature T-cells
3.	**Lymph Node**		
	Para-cortex	Inter-digitating cells, T-helper cells, T-suppressor cells	Lymphocyte entry through endothelial venules, antigen presentation to T helper cells, T-cell proliferation-differentiation-regulation
	Primary and Secondary follicle	Dendritic cells, dendritic macrophages, B-cells, and small no. of T cells	Storage of (Naïve/memory) B-cells, re-circulating B cells (IgM^+, IgD^+)
	Germinal center	Follicular dendritic cells, dendritic macrophages, B -cells, helper T-cells	T-cell dependent B lymphocytes, differentiation, antigen presentation
	Medulla	Reticular cells, T-effector cells, plasma cells, polymorphonuclear granulocytes	Termination of antigen specific reaction, antibody synthesis and immune-complex mediated clearance, T-delayed type hypersensitivity reaction and cytotoxic cells.

Sl.No.	*Lymphoid Organ and Compartments*	*Types of Cells Present*	*Functions*
4.	**Spleen**		
	Red pulp	Reticular cells, T-effector cells, plasma cells, polymorphonuclear granulocytes	Termination of antigen specific reaction, antibody synthesis and immune-complex mediated clearance, T-delayed type hypersensitivity reaction and cytotoxic cells.
	Marginal Zone	Macrophages, marginal metallophilic cell	T-cell independent B lymphocyte proliferation-differentiation, B- cell memory
	White pulp	Plasma cells, helper T-lymphocytes, B-lymphocytes	Site of lymphocyte production
5.	**Organized Lymphoid Tissues**		
	MALT, NALT, BALT, Peyer's Patches	B and T lymphocytes, dendritic cell, plasma cells	Initiation of immune responses including IgA response, uptake of exogenous substances
6.	**Single Cells**		
	Epithelium	Epithelial cells, T-cytotoxic cells, natural killer cells,	First line of defense, extra-thymic maturation of lymphocytes, transport of IgA to lumen
	Lamina Propria	Plasma cells, macrophages, lymphocytes	Synthesis of IgA antibody, phagocytosis and killing

Source: Kuper *et al.*, 2000.

- ☆ Forceps
- ☆ Pins
- ☆ Brush
- ☆ Wash bottle
- ☆ Flag labeling strips

4. Procedure

4.1. Opening the Animal

1. Take a clean dissecting tray and put the rat on its back.
2. Stretch the limbs and the fix them with the pins. Stretch the snout upwards and fix it with a pin.
3. With the help of a pair forceps lift the skin of the abdomen from the middle and make a cut with the help of sharp scissors.
4. Extend the cut vertically upward from the abdomen to the snout and downwards till the genital opening.
5. The skin of the four limbs is separated by making a transverse incision. With the help of a scalpel, separate the skin and fix them in the tray with the help of pins.
6. Cut the abdominal wall at the posterior region. Continue the incision anteriorly. Cut the ribs up to anterior end of the thoracic cavity. Also make lateral incision to open the animal properly.
7. Carefully cut the clavicles.
8. Proceed further to the anterior end, cut the connection of the diaphragm and the abdominal wall.
9. Separate the ventral wall of the abdomen and the thorax and pin the skin flaps to expose the internal organs.
10. Remove the fat depositions with the help of a brush by washing it with water.
11. Clean the tray and observe the anatomy of the animal (Figure 1b).

4.2. Display of Lymphoid Organs and Flag Labelling

Only the large lymphatic organs spleen and the thymus glands are visible in the dissected animal. On the inner surface of the skin, many lymph nodes are also visible.

1. Take a strip of thick white sheet and insert below the spleen and thymus gland.
2. Observe the lymph nodes carefully and mark them.
3. Write the names of these organs on different strips and fix them on a long needle like a flag.

4. Pin the flag on the tray in front of the respective organ to complete the display.

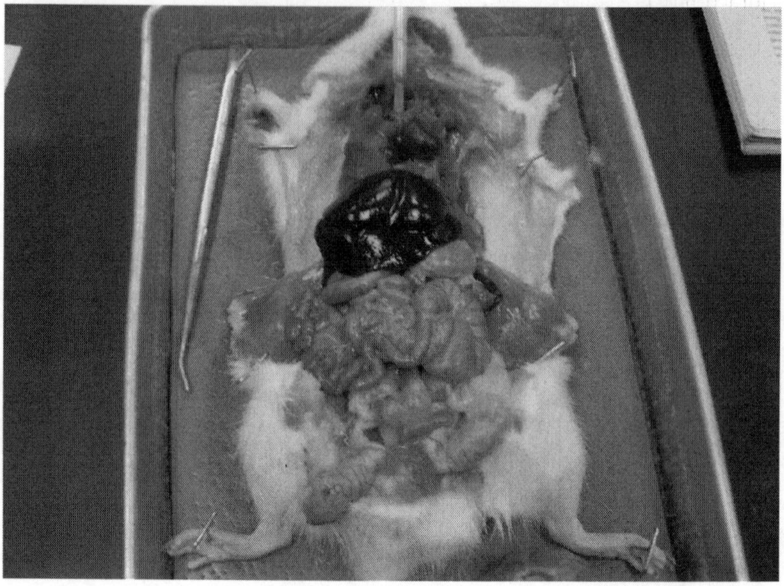

Figure 1b. Display of Anatomy of Dissected Rat.
(*Source*: By Allen Lew from Berkeley, USA - Flickr, CC BY 2.0).

5. Observation and Discussion

Apart from the classification of the lymphoid organs as primary and secondary organs, they are also classified as **external and internal lymphoid organs** based on their location. *Thymus, spleen, and lymph nodes are the internal lymphoid* **organs.** They are specialized organs taking part in the immune mechanism and can be easily seen in a dissected rat. *External lymphoid tissues such as MALT, NALT, SALT, Peyer's patches,* are aggregates of lymphoid follicles localized in different parts of the body. For example, skin does not contain organized lymphoid tissues, however, its immune components are connected with other lymphoid organs and constitute the Skin-associated lymphoid tissue (SALT). These structures are scattered and located deep inside different organs of our body such as nose, lungs, gastrointestinal tract, reproductive organs *etc.*, and cannot be seen in a dissected animal.

Visualizing all lymphoid organs in a dissected animal is one of the most challenging of anatomic dissection. The position and anatomy of these organs make it difficult to be located, in a dissected animal. Thymus and Spleen are large organs and can be seen easily. Tonsils are large enough to be visible but their location in the laryngopharyngeal region makes the display difficult. *Rats and mice lack tonsils.* Narrow transparent channels of lymph vessels are present beneath the skin surface and the gastro-intestinal tract. These vessels transport the lymphatic fluid to the lymph nodes. Few large lymph nodes are seen below the surface of the skin. Other lymphoid structures such as Peyer's patches in intestinal wall and bone marrow are not visible in a simple anatomical display (Figure 1c).

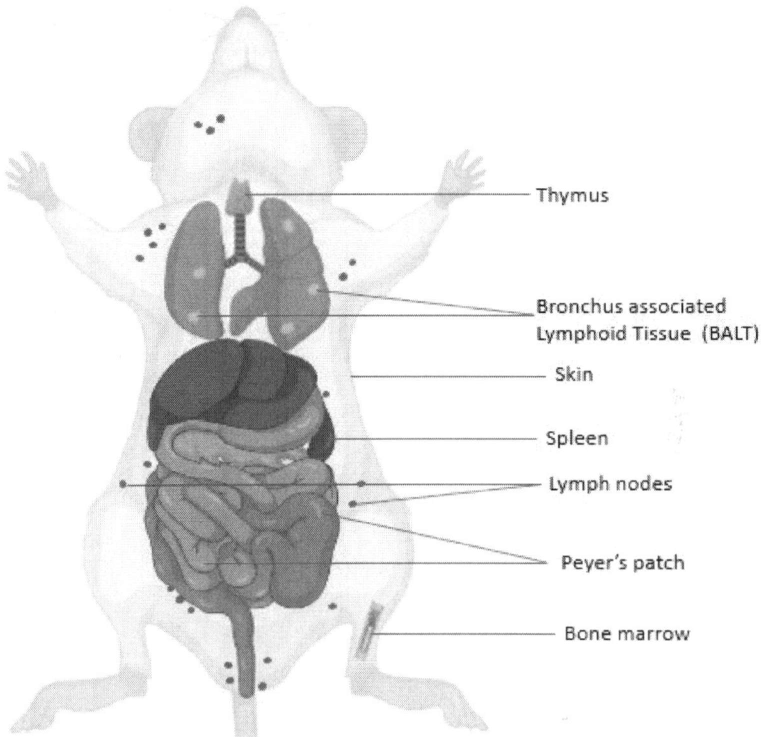

Figure 1c. Schematic Representation of Lymphoid Tissues in Rat.

The major lymphatic organs present in a rat are (Figure 1d):

Thymus

Thymus is the primary lymphoid organ present in mammals. Thymus sits above the heart along the midline of the thoracic cavity, in the anterior part of the mediastinum, behind the sternum. It is a bilobed milky white gland. The size of thymus is larger in younger rats but may be difficult to find in older animals. Lymphocytes produced in the thymus are known as T-lymphocytes. The T cell mature and differentiate in thymus. Thymosin hormone is secreted by thymus gland.

Spleen

Spleen is a largest secondary lymphoid organ. It is located behind the stomach in the abdomen and entirely surrounded by peritoneum. It is present in the peritoneal cavity, on the left side, high under the rib cage, close to the diaphragm. It is soft and dark red in color as it is a highly vascular organ. It is a major site for the antibodies and effector T-cells. It is made up of two types of tissues: the red pulp and the white pulp tissues. In mice, spleen is the reservoir of more than 50 per cent of the monocytes. Upon injury, the monocytes of the spleen migrate to the site of injury and transform into the dendritic cells and macrophages and help in the repair of the injury. Spleen is also known as the graveyard of RBCs.

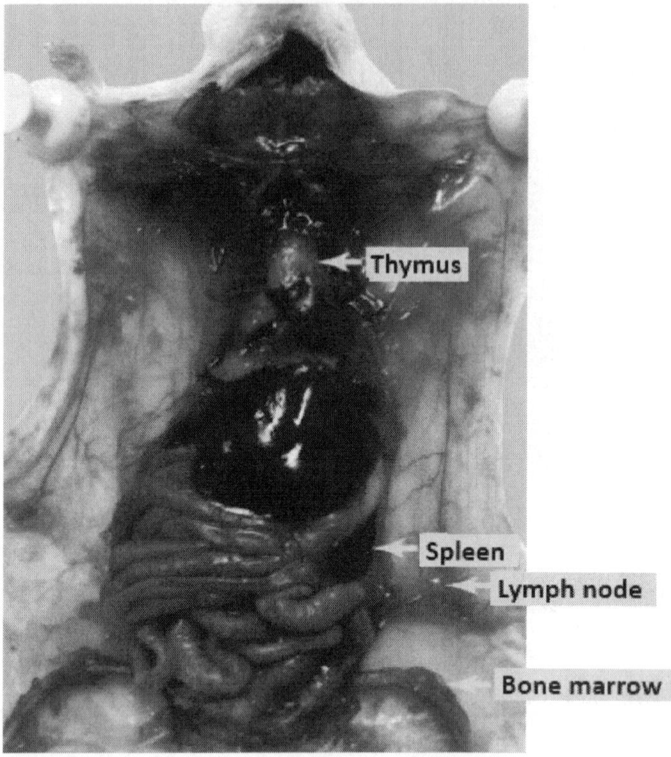

Figure 1d. Display of Lymphoid Organs in a Dissected Rat (*Source*: Mak *et al.,* 2014).

Lymph Nodes

Lymph node is a secondary lymphoid organ of the lymphatic system and is also referred as lymph gland. The lymph nodes are rounded, or bean shaped structures placed on the lymphatic vessels in such a way that they can filter out any foreign material carried in the lymph. Lymph nodes are difficult to find in a dissected rat. Lymph nodes appear as white pearly bead like translucent structures that easily blends with the fat deposits. The most prominent nodes are the *brachial, mesenteric, and inguinal nodes.* Most of the lymph nodes are small and difficult to locate. Injectable dyes and stains are used to study the network of lymphatic node in animals.

Peyer's Patch

Peyer's patches are the areas of lymphoid tissues located in the intestinal walls. They appear as elongated thickenings of the intestinal epithelium but not very distinctly observed in a dissected rat. They are the secondary lymphoid organs and contain variety of immune cells such as macrophages, dendritic cells, lymphocyte and microfold cells (M cells). It serve as the site of antigen interaction and lymphocyte activation. The jejunum and the ileum region have greater number of patches compared to that in the duodenal region. These patches are not visible in the dissected rats.

Bone Marrow

Bone marrow is the largest mass of primary lymphoid tissue in the body. It is a semi-solid tissue found in the spongy part of the bones. It is also the primary site of hematopoiesis. It comprises of the hematopoietic stem cells, adipose tissues, and the stroma cells. Hematopoiesis gives rise to all the three types of blood cells, *i.e.*, erythrocytes (RBC), leukocytes (WBC), and thrombocytes (platelets). During a primary immune response, the antibodies produced in the body are derived only from the spleen and the lymph nodes. Towards the end of the response, the memory cells leave the spleen and colonize the bone marrow. After a second exposure to the same antigen, the bone marrow produces large quantities of antibody which contributes to 70 per cent of the total antibodies produced.

NALT, SALT, BALT, MALT

The nasal associated lymphoid tissues (NALT) are the lymphatic aggregates present in the nose, skin associated lymphoid tissue (SALT) are aggregates of lymphocytes on the surface of skin, bronchus associated lymphatic tissues (BALT) are present throughout the lungs, and MALT -Mucous associated lymphocyte tissues are present in the gastrointestinal region. These are not visible in the dissected rat.

6. Precautions

- ☆ Do not make deep cut on the skin. Make sure that the tips of scissors point upwards while making the incision.
- ☆ Remove the layer of skin with the help of scalpel till the end of the extremities to expose the organs properly.
- ☆ Avoid damaging the underlying organs and blood vessels.
- ☆ Remove the excess fat if needed for a proper visualization.
- ☆ Rinse the animal with water and discard the waste once the body cavity is opened.

7. Questions

Multiple Choice Questions

1. Which of the following is NOT a secondary lymphoid organ?

a. Thymus b. Spleen

c. Mesenteric lymph node d. Peyer's patch

2. **Which of the following statement is true for T lymphocytes?**
   a. They can kill and ingest pathogens
   b. They recirculate through the blood and lymphoid organs
   c. They are short lived.
   d. They differentiate in the bone marrow
3. **Which of the following is an external lymphoid organ?**
   a. Thymus b. Spleen
   c. Lymph nodes d. MALT
4. **Peyer's patches are present in**
   a. Lungs b. Skin
   c. Stomach d. Intestine
5. **Which of the following cannot be observed in a dissected rat?**
   a. Tonsils b. Spleen
   c. Lymph node d. Thymus

Short Answer Questions

1. **What is the difference between primary and the secondary lymphoid organs?**
2. **Which is the largest primary and the secondary lymphoid organ.**
3. **Differentiate between the T-lymphocytes and the B-lymphocytes.**
4. **Why are the tonsils not visible in a dissected rat?**
5. **From which cells do the lymphoid cells originate?**

REFERENCES

By Allen Lew from Berkeley, USA - Flickr, CC BY 2.0, https: //commons.wikimedia.org/w/index.php?curid=2277548

Eroschenko, V. P., and Di Fiore, M. S. (2013). *DiFiore's atlas of histology with functional correlations.* Lippincott Williams and Wilkins.

https: //commons.wikimedia.org/wiki/File: Anatomy_of_the_lymphatic_system.jpg

Kuper, C. F., Harleman, J. H., Richter-Reichelm, H. B., and Vos, J. G. (2000). Histopathologic approaches to detect changes indicative of immunotoxicity. *Toxicologic pathology, 28*(3), 454–466. https: //doi.org/10.1177/019262330002800317

Mak, T. W., Saunders, M. E., and Jett, B. D. (Eds.). (2014). Chapter 2—Components of the Immune System. In *Primer to the Immune Response (Second Edition)* (Second Edition, pp. 21–54). Academic Cell. https: //doi.org/10.1016/B978-0-12-385245-8.00002-9

2

Histological Study of Spleen, Thymus and Lymph Nodes through Permanent Slides/Micrographs

Dr. Sushma Bhardwaj

ABSTRACT

The lymphoid system consists of all the cells, tissues and organs that contain accumulations of the immune cells called the lymphocytes. The lymphocytes are distributed throughout the body in various forms:

- ☆ *As single cells,*
- ☆ *Isolated group of cells,*
- ☆ *As distinct non-capsulated lymphatic nodules present in the connective tissues of the digestive system, respiratory system or reproductive systems (e.g., peyer's patch)*
- ☆ *As encapsulated individual lymphoid organs (e.g., lymph nodes, spleen)*

This section will deal with the histological study of the major lymphoid organs of immune system, i.e., lymph nodes, thymus, spleen and tonsils.

***Keywords**: Lymphoid organs, Lymph node, Spleen, Thymus.*

Learning Objectives

After performing this experiment, students will get an overview of the lymphoid organs and the associated terms. They will be able to:

- ✰ Locate the position of major lymphoid organs
- ✰ List the morphological structure and the functions of lymphoid organs
- ✰ Understand the histology of spleen, thymus, lymph node and tonsil
- ✰ Differentiate between the various lymphoid organs based on their structure, histology and functions.

1. Introduction

Lymph nodes, spleen, tonsils, and thymus are the major organs of immune system. Lymph nodes and spleen are the encapsulated lymphoid organs in mammals.

The **lymph nodes** are widely distributed along the path of the lymphatic vessels. The most prominent nodes are found in the axillary and the inguinal regions. A capsule of connective tissues is present around the lymph nodes from which projections called the trabeculae are sent inwards. The nodes are divided into cortex and medulla containing light staining central regions called the germinal centers in the cortical region and the medullary cords and sinuses in the medullary regions. The afferent lymphatic vessels enter the lymph node which pass through the cortex and medullary regions and exit via the efferent lymphatic vessels.

Spleen is the largest lymphoid organ. It is highly vascular. The trabeculae of the capsule divide the interior into incomplete chambers called splenic pulp, containing white and red pulp. White pulp consists of darkly stained lymphatic nodules surrounded by a central artery. Red pulp surrounds the white pulp. It consists of splenic sinusoids and splenic cords which is formed of network of reticular fibers containing macrophages, lymphocytes, plasma cells and blood cells. Splenic sinuses are the network of blood channels which collect the blood in larger sinuses and finally leaves the spleen via splenic vein.

Thymus is a soft lobular gland present in the lower part of the neck, in the anterior mediastinum region. It is the site of maturation of the T-lymphocytes. It is the most important and active organ of immunogenesis during childhood. This is an exceptional lymphoid organ with some unique properties. Unlike other lymphatic organs which are strategically positioned to encounter the antigen, thymus is protected from the antigenic exposure. Also, the mitotic rate is higher than other lymphatic tissues, though the number of cells that leave the gland is far less. This is because, a large number of lymphocytes produced in thymus die as a mechanism to remove the "forbidden' autoreactive lymphocytes against the '*self-antigens*'. Like the lymph nodes and the spleen, thymus too is surrounded by capsule of connective tissues. Beneath the capsule is the darkly stained cortex with a network on interconnected spaces which harbor the immature lymphocytes. The lightly stained region is the medulla having coarser framework of epithelial cells which contain fewer lymphocytes. Whorls of the epithelial cells of the medullary regions form *thymic (Hassall) corpuscles.*

Tonsil is the aggregate of lymphatic tissue located in the wall of the nasopharyngeal region, anterior to the throat of humans and other mammals. Rats and mice lack tonsil, and the function is performed by the NALT. In man, three sets of tonsils are present which are also referred as *palatine tonsils.* Tonsils are a pair of oval masses that protrude from the sides of the oropharyngeal wall behind the oral cavity. The exposed surface of the tonsil has many pits which lead the lymphatic tissues present beneath these lymphatic aggregates. When the debris lodges in the pits and chokes them, it causes inflammation. This condition is called *tonsilitis,* which is a common nasal inflammation seen in man.

2. Requirement

i. Permanent slides of T.S through spleen, thymus and lymph node, palatine tonsil
ii. Binocular microscope (4× and 10× magnification)

3. Observation

The slides are put under the microscope and the section is observed under 4×, 10× and if needed under the 40× magnification. The histology of the tissues under the microscope was studied by observing the different regions, and types of cells that are present in the tissue.

3.1. T.S. Lymph Node

3.1.1. Histological Features

Panoramic View of Lymph Node: 10× view: (Figure 2a)

The panoramic view of lymph node shows the darkly stained region of cortex and the light stained medulla, with the following features:

1. The outer capsule is surrounded by the perivascular adipose tissues which are highly vascular.
2. The capsule is made of connective tissues which are densely organized and surround the lymph node.
3. The capsule extents inward as trabeculae, pass through the cortex and ramify in the medulla.
4. Connective tissues of the trabeculae contain the major blood vessels.
5. The afferent lymphatic vessels penetrate the capsule and enter cortical sinus in the cortex of the lymph node.
6. The cortical sinuses extend into the medulla along with the trabeculae and form the medullary sinuses.

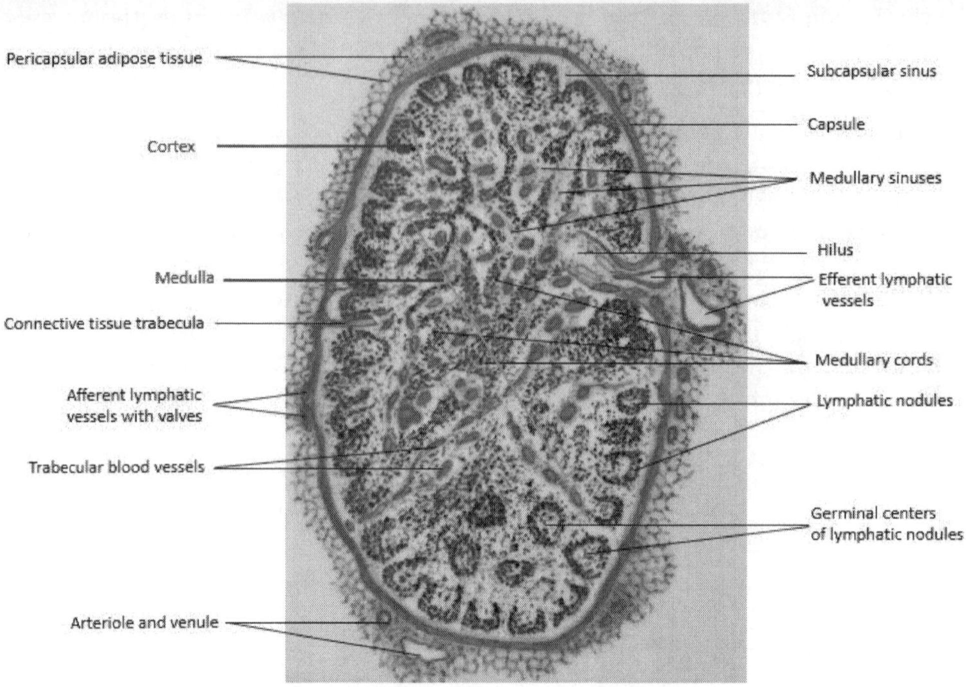

Figure 2a. T. S. of Lymph Node (Panoramic view): Stain: Hematoxylin and eosin (10×).

7. In the cortex, large number of lymphatic nodules are seen. The section through the lymphatic nodules exposes the lightly stained areas called germinal centers. The germinal centers are the sites where the lymphocytes proliferate.
8. In the medulla of lymph nodes, lymphocytes aggregate to form irregular chords of lymphatic tissues called the medullary cords which contain small lymphocytes, macrophages, and plasma cells.
9. The medullary sinuses drain the lymph received from the cortical region.
10. The lymph passes through the medullary cords and collect in the hilus of the lymph node which is a concave depression in the node where the nerve, blood vessels and lymph vessels are located.
11. The efferent lymphatic vessels collect the lymph from the medullary sinus and leave the lymph node in hilus.

T.S. of Lymph Node through Cortex and Medulla: 10× view (Figure 2b)

The section shows the following features:

1. Capsule is surrounded by the connective tissues which has a venule and an artery.

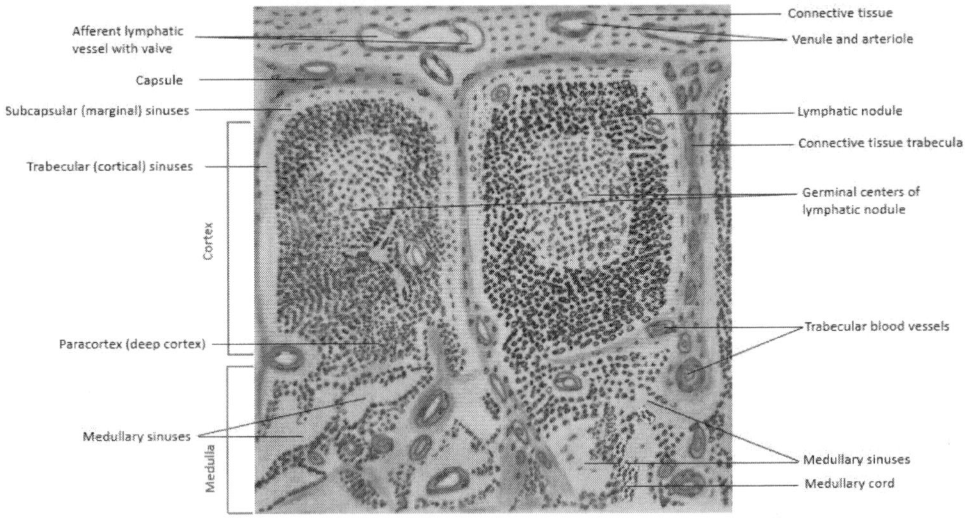

Figure 2b. T.S. of Lymph Node:
Stain: hematoxylin and eosin. (40× view).

2. Afferent lymphatic vessel with endothelial lining is present in the connective tissue layer.
3. Trabeculae arise from the inner surface of the capsule which extend through the cortex and medulla.
4. Sub-capsular or marginal sinus is seen between the capsule and the cortex.
5. Cortex consists of lymphatic nodules which are arranged adjacent to each other partially separated by the trabeculae and the cortical sinuses.
6. A central light staining germinal center is present in the section through the lymphatic nodules.
7. The cells of the germinal center are loosely arranged. Proliferating lymphocytes have large nuclei and more cytoplasm.
8. The deeper portion cortex is called the paracortex having T-lymphocytes. It is the zone which transits from cortex to the medullary cords of the medulla contain the medullary sinuses
9. Reticular connective tissue network provides the structural support to the cortex and medulla of the lymph node and forms the core.
10. Medullary sinuses have very few lymphocytes, therefore the reticular framework of the connective tissues can be easily distinguished in the nodules and the medullary cords.

3.1.2. Functions of Lymph Node

Lymph nodes are the most important component of immune system. They are distributed throughout the body along the path of the lymphatic vessels and play a major role in the defense mechanism. The prominent functions of lymph nodes are:

1. Filtration of the lymphatic fluid.
2. Phagocytosis of invading bacteria and other substances from the lymph and prohibiting them from reaching the circulatory system.
3. Lymph nodes are the site of production, storage and activation of the T cells and the B cells.
4. Lymphocytes proliferate in lymph nodes and the B cells transform into the plasma cells.
5. Lymphatic fluid that leaves the lymphatic fluid contain large number of antibodies which are distributed along the body.
6. The B cells congregate in the lymphatic nodules present in the outer cortical region of the lymph node, while the paracortex of lymph node is the site of assembly of T cells.
7. Lymph nodes are the site of antigen recognition and activation of B cells resulting in the formation of plasma cells and the B memory cells.
8. After the activation of the B cells, lymphocytes proliferate in the germinal centers of the lymphatic nodules.

3.1.3. Key Identification Features in the Slide

- ☆ The T. S. shows the oval structure with a distinct capsule, cortex and the medullary regions at 4× magnification.
- ☆ The lightly stained lymphatic nodules arranged along the cortex is visible. The germinal centers are seen in the center of the nodules.
- ☆ Trabeculae projecting from the capsule in the cortical and the medullary regions are visible.
- ☆ Cortical and medullary sinuses, hilus and efferent lymphatic vessels are seen in the slide at low magnification.
- ☆ Higher magnification shows the trabecular vessels and medullary cords.

3.2. T.S Spleen

3.2.1. Histological Features

Panoramic View of Spleen 10× view (Figure 2c)

The panoramic view of spleen shows the following features under low magnification:

1. A capsule of densely packed connective tissues encloses the spleen from which the trabeculae arise which extend towards the interior.
2. Trabecular arteries and veins are present inside the trabeculae.

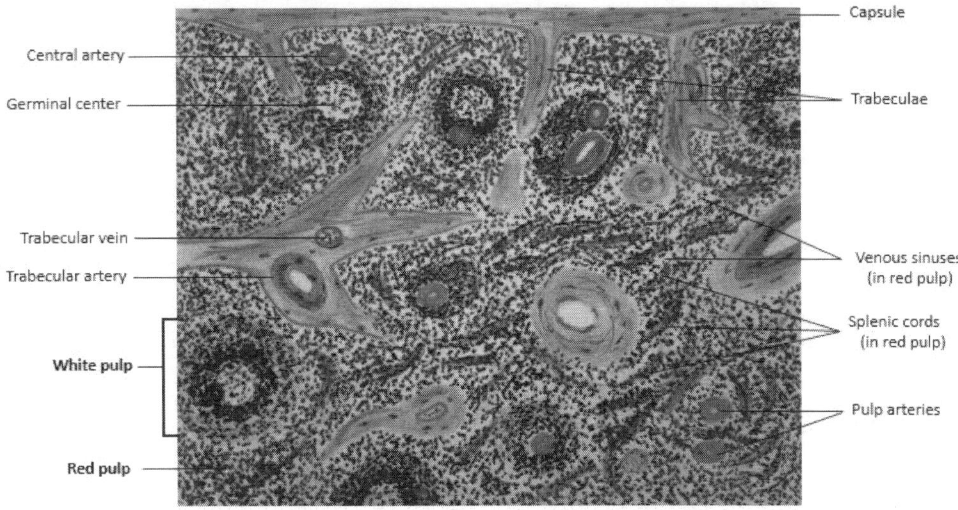

Figure 2c. T.S. of Spleen (Panoramic View). Stain: Hematoxylin and eosin. (10×).

3. The trabecular arteries leave the trabeculae and gets unsheathed with the lymphatic tissue and branch off as many central arteries that pass through the lymphatic nodules.
4. Large number of lobular aggregates of lymphatic nodules are seen throughout the spleen which constitute the white pulp of the spleen. Lymphatic nodules have the germinal centers.
5. A central artery passes through the periphery of each lymphatic nodule, which branches off from the trabecular artery.
6. The lymphatic nodules are surrounded by the cellular meshwork diffused with the connective tissue trabeculae called the splenic pulp commonly known as red pup as it contains extensive vascular tissue.
7. Red pulp contains pulp arteries, splenic chords of 'Billroth' and splenic sinuses.
8. Splenic cords are diffused strands of lymphatic tissues that form spongy mesh of reticular connective tissue.
9. In addition to the splenic sinuses, spleen also contains venous sinuses.
10. The capsule and the trabeculae in the spleen are thicker than those found in the lymph nodes.

T.S of Spleen through Red and White Pulp: 40× view (Figure 2d)

An elaborate histological arrangement of the red and white pulp, associated trabeculae, blood vessels and sinuses are visible at higher magnification of the section of spleen.

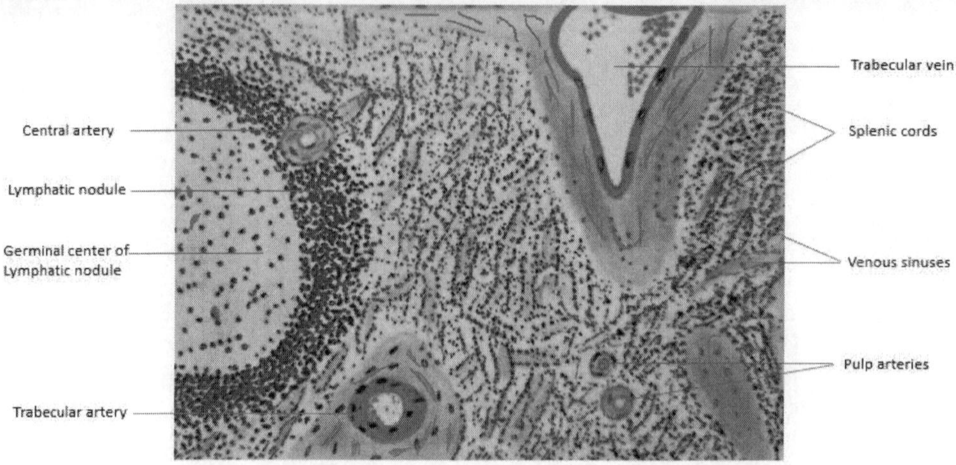

Figure 2d. T.S. of Spleen through Red and White Pulp. Stain: Hematoxylin and eosin. (40×).

1. Large lymphatic nodules are present throughout the spleen which represent the white pulp.
2. Each lymphatic nodule or white pulp has
   - ✰ Peripheral zone – It is the peri-arterial lymphatic sheath made up of densely packed small lymphocytes. The cells of this sheath are T lymphocytes.
   - ✰ Central artery – It is eccentric in position with respect to the lymphatic nodule but present in the center of the peripheral zone, therefore called the central artery.
   - ✰ Germinal center – This may or may not be present. The germinal center decreases in number with age. They are lightly stained regions of the lymphatic nodule and contain the B lymphocytes.
3. Red pulp contains
   - ✰ The splenic cords (of Billroth) - These are formed by the aggregation of small lymphocytes, associated cells and the blood vessels.
   - ✰ Venous sinuses – Splenic cords are interrupted by the venous sinuses. These are dilated blood vessels which are lined by elongated endothelial cells.
   - ✰ Pulp arteries – After leaving the lymphatic nodules, the central artery branches into pulp arteries. Capillaries and pulp veins are also seen in the red pulp.
4. Connective tissue trabeculae are prominent features of the section apart from the red and the white pulp.
5. Trabecular artery and the veins are seen inside the trabeculae.

3.2.2. Functions of Spleen

1. Spleen is the largest lymphoid organ having an extensive blood supply.
2. It filters blood and is also the site of immune response to the antigens present in blood.
3. Red pulp is rich in erythrocytes, lymphocytes, granulocytes, plasma cells and the macrophages. It filters the blood by removing the aged or abnormal erythrocytes, platelets, microorganisms, and other antigens.
4. The white pulp is the immune component of the spleen. It harbors the T cells in the peri-arteriolar lymphatic sheaths and B lymphocytes in the lymphatic nodules.
5. Antigen processing cells (APCs) and macrophages are also present in the white pulp which detects the trapped antigen and initiate the immune responses against them.
6. These are the site of the activation and interaction, and proliferation of the T cells the B cells.
7. Macrophages of spleen breakdown the hemoglobin of RBCs that are damaged and worn-out. Iron from the heme, is recycled and returned to the born marrow for erythropoiesis.
8. Spleen functions as the hematopoietic organ during the fetal life. This function, however, ceases after birth.
9. Spleen also acts as the reservoir of blood because of its sponge like micro-structure. The stored blood is reverted to the circulation when required.

3.2.3. Key Identification Features in the Slide

- ☆ The T.S of spleen shows the characteristic white and red pulp.
- ☆ Capsule and the trabeculae containing the blood vessels are clearly seen in the slide.
- ☆ The trabecular arteries and veins are visible in the section at high magnification.
- ☆ White pulps with germinal centers are the distinctive features in the T.S of spleen.

3.3. T. S. Thymus Gland

3.3.1. Histological Features

The histology of the thymus gland changes with age. A fully active and fully developed thymus gland is present after birth. As the individuals reach puberty, thymus involute and starts degenerating resulting in a decline in the lymphocyte production.

Panoramic View of Thymus Gland: 10× View (Figure 2e)

The thymus gland is a lobulated lymphoid organ enclosed in a capsule. The panoramic view of the T.S. of thymus gland shows the following features:

1. The thymus gland is enclosed in the capsule formed by the dense connective tissues.
2. From the inner side of the capsule, connective tissue trabeculae arise which extend into the interior and divides the gland into many incomplete lobules.
3. The lobules are divided into darkly stained outer cortex and lightly stained medulla at the inner side.
4. The blood vessels enter the thymus gland passing through the capsule via trabeculae.
5. Inside lobular cortex has densely packed lymphocytes are present, while medulla has very few lymphocytes.
6. Medulla contains many thymic (Hassall) corpuscles which are abundant in the thymus of adults.

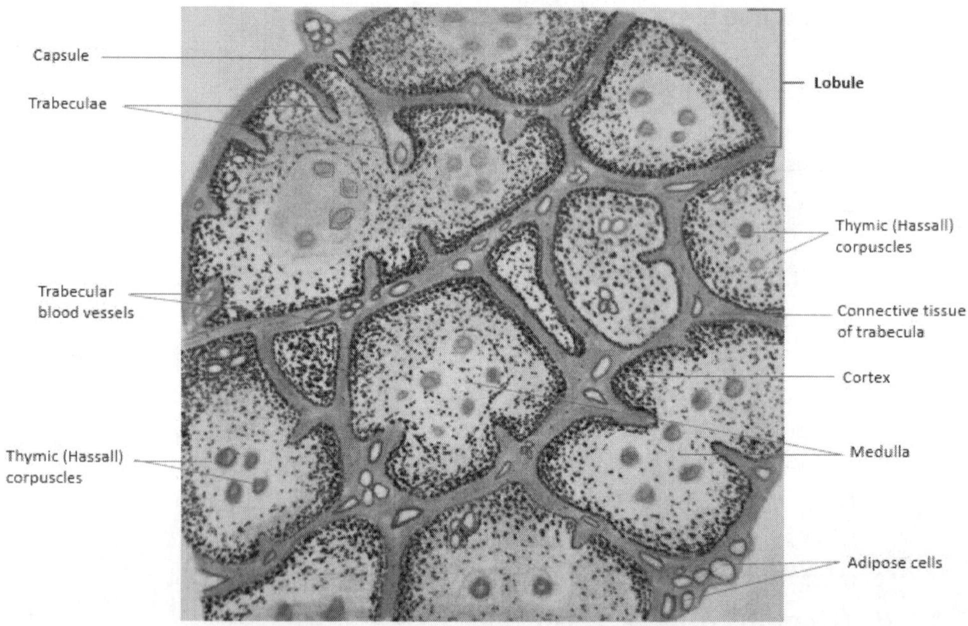

Figure 2e. T. S. of Thymus Gland (Panoramic View), Stain: Hematoxylin and eosin (10×).

T.S of Thymus through Cortex and Medulla: 40× (Figure 2f)

A magnified view through the cortex and medulla of the lobules of thymus shows the following characteristics:

1. Dense aggregation of the lymphocytes is seen in the cortex.
2. Medulla contains lesser number of lymphocytes but many epithelial reticular cells.
3. Oval structures called the thymic or Hassall corpuscles are seen in the medullary region, which are the aggregates of flattened epithelial cells
4. The center of the thymic corpuscles contains the degeneration centers or the calcification centers with unknown function.
5. The thymic lobules and the trabeculae contain the blood vessels and the adipose cells.

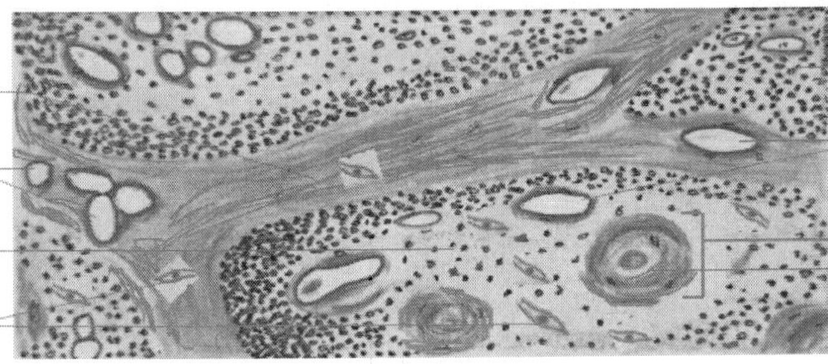

Figure 2f. T. S. of Thymus Gland.
Stain: Hematoxylin and eosin (40×).

3.3.2. Functions of Thymus Gland

Thymus gland is an important lymphoid organ during early development. It has numerous roles:

1. Thymus glands produce a diverse group of T cells.
2. Undifferentiated lymphocytes reach the thymus glands from the bone marrow via blood stream. The reticulo-epithelial cells of the cortex (also called thymic nurse cells) surround the undifferentiated lymphocytes and prompt them to proliferate, differentiate, and mature into helper T cells, cytotoxin T cells and the immunocompetent T cells.
3. The developing lymphocytes are protected against the antigens present in the blood by the blood-thymus barrier which is formed by the macrophages, reticulo-epithelial cells and the endothelial cells.
4. The epithelial reticular cells also secrete the hormones thymulin, thymopoietin, thymosin, thymic humoral factor, interleukins, and interferon for the proliferation, differentiation and maturation.
5. Removal of thymus gland from a newborn baby will result in the production of immune-incompetent T cells. The individual will lack the ability to fight pathogens due to absence of immunological competence.

3.3.3. Key Identification Features in the Slide

- ✰ Lower magnification of the section shows the outer capsule, connective tissue trabeculae and the partially separated lobules.
- ✰ Each lobule has cortex and medulla. The medulla of one lobule is fused with the neighboring lobule as the two lobules are partially separated.
- ✰ Characteristic thymic (Hassall) corpuscles are distinctly seen in the slide.

3.4. T. S. Palatine Tonsil

A pair of tonsils are present in the lateral oropharynx which contain the aggregates of lymphatic nodules.

3.4.1. Histological Features

Palatine tonsils, commonly known as tonsils, are present in man but not in rats and mice. In mice their functional equivalent is known as NALT which are the associated lymphoid tissues of nose. The NALT in mice is formed by a pair of lymphoid aggregates in nose, whereas in man, it is formed of a number of mucosa associated lymphoid follicles.

T.S. of Palatine Tonsil: 10× (Figure 2g)

T.S of the Palatine tonsil show the following histological features:

1. The tonsils are covered by a protective layer made of stratified squamous non-keratinized epithelial cells which covers the entire oral cavity.
2. Deep grooves called tonsillar crypts invaginate each tonsil which are also lined by the stratified squamous non-keratinized epithelium.

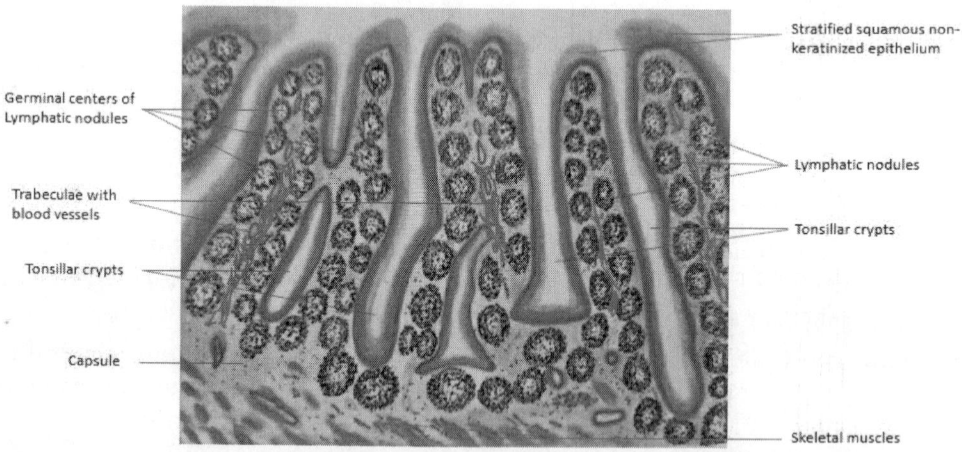

Figure 2g. T.S. of Palatine Tonsil.
Stain: hematoxylin and eosin (10× view).

3. Numerous lymphatic nodules are present beneath the epithelial layer which are distributed along the tonsillar crypts.
4. The lymphatic nodules merge at many regions along the length of the crypts. Germinal centers are visible in the nodules.
5. Tonsils are underlined by thin layer of dense connective tissue capsule which gives rise to the connective tissue trabeculae containing blood vessels. The trabeculae pass the tonsil between the lymphatic nodules.
6. A layer of skeletal muscles is present below the connective tissue capsule.

3.4.2. Functions

1. Palatine tonsils prevent the infection in the respiratory and digestive tracts.
2. The B cells and T cells proliferate and mature in the lymphatic nodules present along the tonsillar crypts.
3. B cells of the palatine tonsils produce all five major immunoglobulin classes.
4. B cells are also found to produce antibodies against many bacteria and viruses.
5. Considerable T cell responses are also initiated in the palatine tonsils following the antigenic stimulation.

3.4.3. Key Identification Features

- ☆ Presence of deep tonsillar crypts
- ☆ Presence of lymphatic nodules with germinal centers.
- ☆ Skeletal muscle layer underlines the tonsil.

4. Questions

Multiple Choice Questions

1. Encapsulated individual lymphoid organs distributed across different parts of the body is called

a. Lymph node
b. Peyer's patch
c. MALT
d. Tonsils

2. The white pulp of the spleen is formed of

a. Trabecular connective tissue
b. B and T lymphocytes
c. Granulocytes
d. Erythrocytes

3. What will happen if the thymus gland is removed from a newborn baby?

a. Infection in the respiratory and digestive tracts
b. Production of incompetent T cells

c. Total absence of T cells

d. Absence of B cells

4. Tonsil is absent in rat. Its functional equivalent present in them is

a. MALT

b. SALT

c. GALT

d. NALT

5. Which of the following will not be seen in the T.S of a lymph node?

a. Germinal center

b. Capsule

c. Trabecular connective tissue

d. Hassall corpuscle

Short Answer Questions

1. **Name the types of cells present in the cortex and medulla of lymph nodes.**
2. **How will you identify the red and the white pulp of spleen?**
3. **Name the hormone secreted by the thymus gland.**
4. **What is the difference between the NALT of mice and man?**
5. **List the function of spleen.**
6. **Name the lymphatic vessels which enter and exit the lymph nodes.**
7. **Where are the Hassall corpuscles found?**

REFERENCES

By Allen Lew from Berkeley, USA - Flickr, CC BY 2.0, https: //commons.wikimedia.org/w/index.php?curid=2277548

Eroschenko, V. P., and Di Fiore, M. S. (2013). *DiFiore's atlas of histology with functional correlations*. Lippincott Williams and Wilkins.

https: //commons.wikimedia.org/wiki/File: Anatomy_of_the_lymphatic_system.jpg

Kuper, C. F., Harleman, J. H., Richter-Reichelm, H. B., and Vos, J. G. (2000). Histopathologic approaches to detect changes indicative of immunotoxicity. *Toxicologic pathology, 28*(3), 454–466. https: //doi.org/10.1177/019262330002800317

Mak, T. W., Saunders, M. E., and Jett, B. D. (Eds.). (2014). Chapter 2—Components of the Immune System. In *Primer to the Immune Response (Second Edition)* (Second Edition, pp. 21–54). Academic Cell. https: //doi.org/10.1016/B978-0-12-385245-8.00002-9

3

Preparation of a Stained Blood Film to Study Types of Cells

Dr. Shailesh Kumar R. Yadav

ABSTRACT

A blood smear, also known as a peripheral blood smear or blood film, is a thin layer of blood smeared on a glass microscope slide and then stained to allow microscopical examination of the various blood cells such as white blood cells, red blood cells and platelets. Blood smears are used to investigate haematological (blood) problems and can be used in a variety of situations. It can be used to determine whether a virus or a bacterial infection is present. It is frequently used to diagnose blood parasites in cases of malaria and filariasis or even to detect anaemia, and to determine the source of jaundice. A blood smear is not the same as an automated test like a complete blood count (CBC). Under a microscope, a technician or healthcare provider examines a blood smear and the results and interpretation of a blood smear could provide them with hints to detect a condition.

Keywords: *RBC, WBC, Platelets, Granulocytes, Neutrophils.*

1. Introduction

Blood is a fluid connective tissue, that consists of various specialized cells and is surrounded by a liquid extracellular matrix containing water, salts, nutrients and soluble proteins. The extracellular matrix is called plasma and various cells and cellular fragments are suspended in the liquid matrix.

1.1. Functions of Blood

1. ***Transportation:*** Blood is involved in the transport of gases between the tissues. It carries Oxygen from alveolar tissue to other tissues of the body and carbon dioxide from different tissues of the body to the lungs. It carries nutrients like glucose, amino acids, fatty acids, salts, ions, vitamins and essential trace metals from the gastrointestinal tract to different tissues of the body and hormones from endocrine glands to target tissues. It also transports waste products to excretory organs for elimination from the body.
2. ***Regulation.*** Blood also helps in maintaining the internal environment of the body (homeostasis) in terms of composition, conc., pH, temperature and volume. It helps in maintaining pH by interacting with acids and bases. It also helps in regulating body temperature by absorbing heat and circulating throughout the body and dissipate by varying flow rate through the skin. In addition, the water content inside the cell is maintained by osmotic pressure of blood through interaction with dissolved ions and proteins.
3. ***Protection.*** The clotting mechanism of blood components protects against excessive loss from the cardiovascular system after an injury. In addition, leucocytes present in blood protect against pathogens by destroying them via phagocytosis. Different types of proteins present in blood, antibodies secreted from lymphocytes, interferons *etc.* also help in protecting against a variety of diseases. Natural killer cells help in protecting against cancer cells.

1.2. Composition of Blood

1. Blood plasma: About 55 per cent of total blood content is clear straw-colored fluid portion that constitutes about 91 per cent water and 9 per cent solids (1 per cent inorganic molecules composed of $\mathbf{Na^{+}}$, $\mathbf{Ca^{2+}}$, $\mathbf{Cl^{-}}$, $\mathbf{HCO^{3-}}$ in extracellular matrix, $\mathbf{K^{+}}$, $\mathbf{Mg^{2+}}$, $\mathbf{Cu^{2+}}$, $\mathbf{PO_4^{3-}}$, proteins in intracellular matrix, $\mathbf{Fe^{2+}}$, $\mathbf{Fe^{3+}}$ and **8 per cent** organic molecules composed of plasma proteins and other non-protein nitrogenous substances, sugar, fatty acids, enzymes, vitamins, hormones *etc.* and waste products like creatinine, bilirubin, and urea (Figure 3a)
2. Formed elements: Cells and cell fragments constitute 45 per cent of total blood content and is called packed cell volume (PCV) or hematocrit. It includes:
   (i) Erythrocytes or Red blood corpuscles (RBC)
   (ii) Leucocytes or white blood corpuscles (WBC)
   (iii) Platelets or Thrombocytes

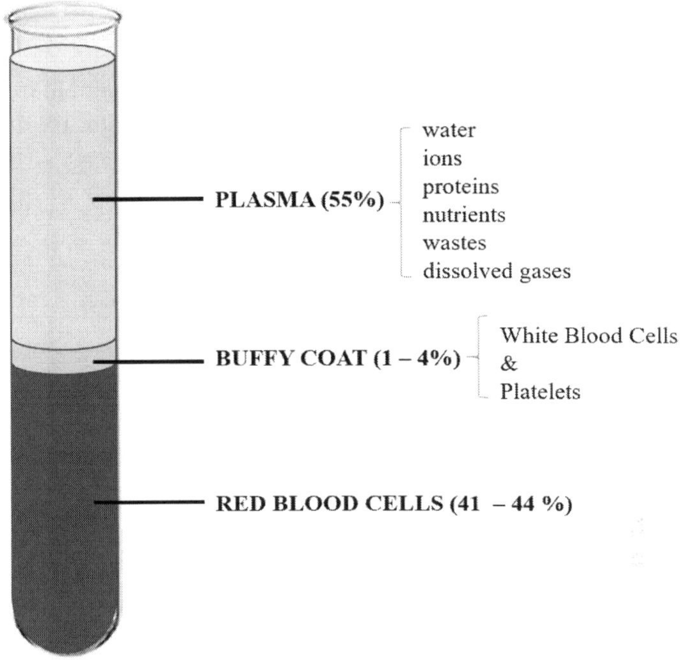

Figure 3a. Composition of Blood- The Plasma and the Formed Elements which Constitute White Blood Cells, Red Blood Cells and Platelets.

(i) Red Blood Corpuscles or Erythrocytes

RBCs are the most abundant cells in the blood and constitutes about 40-45 per cent of total blood. They have a biconcave disc like shape with a mean diameter of 7.8 μm, a thickness of 2.5 μm at the periphery and 1 μm or less in the centre and the average volume of the red blood cell is 90 to 95 μm^3. It is a non-nucleated cell and also lacks other cell organelles during maturation. RBCs can neither undergo mitotic division nor carry other extensive metabolic activities. It contains oxygen-carrying pigment, hemoglobin (about 33 per cent), which consists of pigment 'haem' and polypeptide part 'globin'. The red colour of blood is due to its 'haem' pigment. Its production is controlled by erythropoietin, a hormone secreted from interstitial cells in the peritubular capillary bed of the renal cortex. RBCs have a lifespan of about 120 days and are continuously replaced by new cells because of aging and degradation while squeezing through blood capillaries. Specific inherited surface antigens on the RBC plasma membrane determine blood group. In healthy males, the average number of red blood cells is 5.2 million/μl (±300,000) while in healthy females, it is 4.7 million/μl (±300,000).

Function

- RBCs carry oxygen from the lungs to the different tissues of the body and carbon dioxide from different cells of the body to the lungs for exhalation.

- The free radicals released by RBC after lysis damages bacterial cell wall and plasma membrane.
- Under stress conditions, RBC releases signal molecules such as ATP, S-Nitrosothiols, nitric Oxide, H_2S *etc.*, which promotes blood flow by dilating the blood vessels.

(ii) White Blood Corpuscles (Leucocytes)

Leucocytes are the part of body's protective mechanism. WBCs along with other cells and a variety of plasma proteins constitute the body's immune system, which helps in

- Protecting against pathogenic micro-organism (bacteria, viruses and/or fungi).
- Identify and destroys cancer cells that arise in the body.
- Removes cell debris and worn out cells and helps in tissue repair.
    - Leucocytes constitute about 1 per cent of total blood volume. Total count in the blood is approximately 4500 – 11000/mm^3.
    - They are synthesized in bone marrow hematopoietic stem cells. They travel and enter into almost all parts of the body including the connective tissue, lymphatic system and blood streams.

Diseased Condition Due to Abnormality in WBC

- **Leukopenia** is a condition where the leucocyte count decreases below average level caused by decreased production in bone marrow due to medications (Chloramphenicol and thiouracil), and chemicals (that contain benzene or anthracene nuclei), radiations and/or chemotherapy.
- **Leukocytosis** is a condition where leukocyte count increases in blood above the normal range. It is caused by a number of conditions like inflammatory disease, various types of infections.
- **Leukemia** is a condition where abnormal leukocytes number increase tremendously. Cancerous mutation of myeloid lineage or lymphoid lineage cells, leads to uncontrolled production of abnormal immature leukocytes (Usually >50,000/mm^3).

Leukocytes are further divided on the basis of visible grains present inside the cells, into **Granulocytes** (visible granules present inside the cells) and **Agranulocytes** (no visible grains under the microscope).

There are three types of Granulocytes that are Neutrophils, Eosinophils, Basophils, and two types of Agranulocytes that are Lymphocytes and Monocytes (Figure 3b).

1. Neutrophils

They are the most common white blood cells. The size varies from 10 - 14 µm diameter. The nucleus is multi-lobed (3 – 5), that is why they are also called poly-morphonuclear leucocytes. The cytoplasm contains very fine cytoplasmic granules. The granules are neutrophilic in nature (takes both the acidic and basic stains, therefore, the cells are called as neutrophil. It accounts for 65 - 70 per cent of the total leucocytes count (2000 – 7500 cells/mm^3). Life span of neutrophils varies from 6 hours to few days.

Functions

1. They engulf bacteria through the process of phagocytosis and destroy them intracellularly
2. Neutrophils undergo a specific type of apoptosis called as NETosis, and release granules filled with chemicals, which can kill many bacteria at the same time.
3. They can perform cleaning activity by scavenging cellular debris.

2. *Eosinophils*

Eosinophils constitute about 2 – 4 per cent of total blood leukocytes (40 – 400 cells/mm^3). The diameter varies from 10 - 14µm. The cytoplasm is acidophilic and therefore, they appear reddish pink in colour and have coarse granules. The nucleus is generally 'bi-lobed' and the two lobes are connected with thin chromatin thread give a 'spectacle' like appearance. The life span of eosinophils varies between 8 – 12 days.

Functions

1. Eosinophils bind to larger parasites with special surface molecules and release hydrolyzing enzymes from their granules to kill parasites.
2. Eosinophils also release reactive oxygen species, which is lethal to parasites.
3. Eosinophils secrete granules with 'major basic proteins' which have high larvicidal activity.
4. Eosinophils are mild phagocytic in nature because of restricted motility. They provide immunity to mucosa of respiratory tract, gastrointestinal tract, urinary tract.
5. Eosinophils play role in allergic reactions and limit their intensity by inhibiting mast cell degranulation and the effect of mediators like histamine, bradykinin.

3. *Basophils*

These granulocytes account for about 0.4 – 1.0 per cent (0 - 300 cells/mm^3) of total white blood cells. They have coarser granules in the cytoplasm and stain purple or blue with basic dyes (methylene blue). The pale nucleus is usually hidden by plenty of these granules. The diameter of cells varies between 10 –14 µm. The average life span of basophils varies between a few hours to 3 – 4 days.

Functions

1. During allergic reactions, the immunoglobin E binds to basophils and induces them to release large quantities of histamine, bradykinin, serotonin, heparin and slow reacting substances of anaphylaxis.
2. Basophils secrete anticoagulant heparin which prevents clotting as well as promotes movement of other WBCs.
3. The histamine secreted by basophils in response to allergic reactions dilates the blood vessels and acts as chemoattractant to bring more immune cells to the site of injury/reactions.

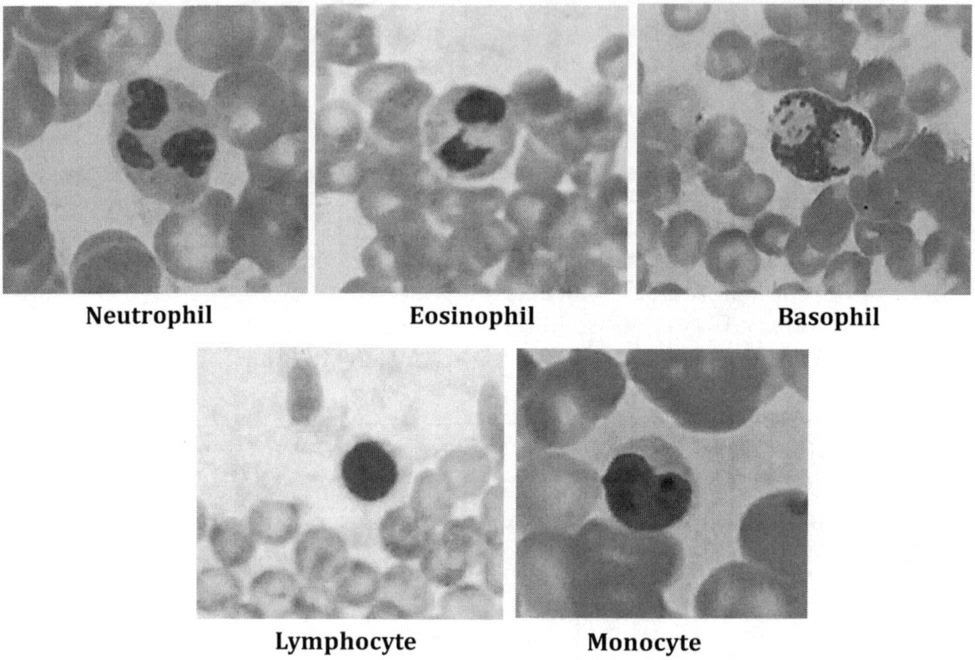

Figure 3b. Different Types of White Blood Cells (Stained with Giemsa stain)*.

***Neutrophil:** Nucleus multilobed (3-5 lobes), very fine cytoplasmic granules; **Eosinophil:** Nucleus bilobed, coarse granules present in cytoplasm; **Basophil:** coarser granules in the cytoplasm and pale nucleus hidden by granules; **Lymphocyte:** small spherical nucleus; and **Monocyte:** Single Bean-shaped or kidney-shaped nucleus.

Lymphocytes

Lymphocytes are agranulocytes. They have a single and very big, rounded or oval nucleus and occupy whole cell leaving marginal cytoplasm around it. They constitute about 20 – 30 per cent (1300 to 4000 cells/mm^3) of total white blood cells. The diameter of lymphocytes varies from 7-8 µm (Small) and 12-15 µm

(Large). The large lymphocytes area precursor of small lymphocytes. The small lymphocytes are responsible for antibody production. Lymphocytes life span varies from about 100 - 300 days. Some memory lymphocytes live for years.

Functions

1. Lymphocytes primarily involve in the body's defense mechanism. There are two types of lymphocytes, B and T lymphocytes (B and T cells). T-cells directly destroy target agents by a mechanism called cell mediated immunity while B -cells secrete antibodies, and are responsible for humoral immunity.
2. They recognize certain antigens on the surface of pathogenic agents (bacteria, viruses *etc.*) and create a memory for the same for future encounter.
3. They recognize and destroy abnormal as well as cancerous cells.
4. They present epitopes (antigenic determinants) on their surface and release certain cytokines to activate different cells of the immune system.

Monocytes

These are the largest white blood cells of the body. They have a single large rounded or kidney shaped nucleus and are present on one side of the cell. The diameter of these cell varies from 15 – 30 µm. They account for about 3 – 8 per cent (200 – 800 cells/mm^3) of total white blood cells. Their life span is from few hours to few days.

Functions

1. Monocytes enter into the area of infections or inflammations and provide a second line of defense after neutrophils. They are active phagocytes.
2. They exit the blood circulation and enter into different tissues, become larger and turn into macrophages.
3. They involve in clearing old, damaged and dead cells of the body. Monocytes may also destroy cancerous cells after sensitization by lymphocytes.
4. They synthesize and secrete complement and other biologically active molecules.

(iii) Platelets (Thrombocytes)

These are the smallest colourless blood cells. They are generally spherical or oval granulated bodies. The nucleus is absent. The average diameter varies from 2 - 5 µm. Leishman staining gives a faint blue colour to the cytoplasm with reddish purple granules. They do not reproduce, there are about 150,000 – 400,000 platelets/mm^3 present in human blood. The life span is about 8 – 12 days.

Functions

1. Platelets play an important role in maintaining hemostasis (spontaneous stoppage of bleeding by a physiological process like platelet adhesion and aggregation at the site of injury.
2. Platelets play a major role in clotting mechanism where the formation of fibrin is involved at site of injury. They secret pro-coagulants (clotting factors) to promote blood clotting.
3. They release nor-epinephrine and 5-Hydroxytryptamine (5–HT) which acts as vasoconstrictor, causing vascular spasms in injured blood vessels.
4. They help in clot retraction by contractile actinomyosin like protein, thrombosthenin, present in platelets pseudopodia.
5. They secrete certain growth factor which is useful in the maintenance of blood vessel linings.

2. Requirements

Glass slides, Cotton swab, Lancet, Light Microscope, Dropper, Giemsa Stain, Spirit.

3. Procedure

1. Take two clean slides.
2. Wipe finger with 70 per cent alcohol and prick it using sterilized lancet.
3. Put a drop of blood at an end of the slide.
4. Take another slide to be used as a spreader for blood drop. Place the edge of the spreader slide just in front of the drop of blood. Make an angle of 45° with blood drop and spread on the slide smoothly forming a thin homogeneous layer of blood smear (as shown in Figure 3c).
5. The smear is allowed to dry followed by a dip in methanol for 1-2 min for fixing the blood cells.
6. Let the slide dry and then stain it with Giemsa stain for 15-20 min.
7. Wash excess stain with tap water and let it air dry.
8. Observe the slide under microscope for blood cells.

4. Observation

The smear of blood shows various types of cells at 40 × magnification. The cells observed are RBC, different types of WBC and thrombocytes.

Different types of cells per cent are shown in table.

Sl.No.	*Type of Cell*	*No. of Cell*	*Per cent constituent*	*Remark*
1.				
2.				
3.				

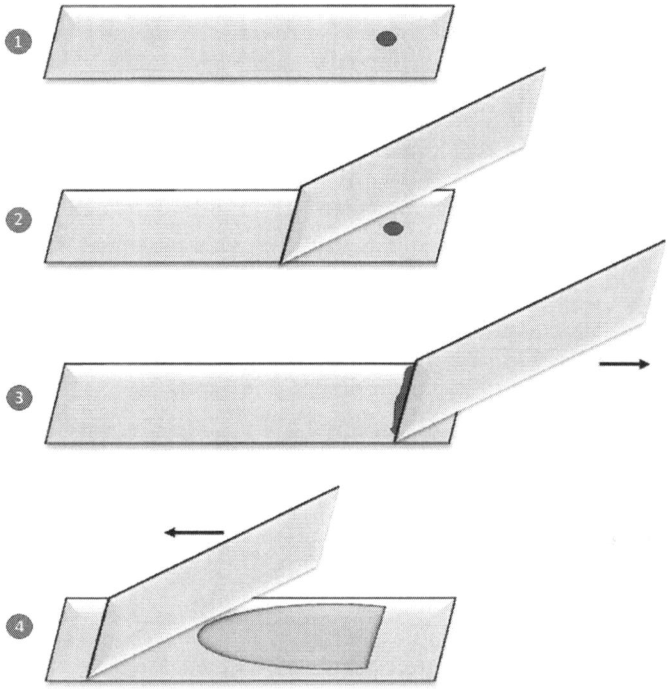

Figure 3c. Preparation of Thin Blood Smear.

1. Put a drop of blood on a clean slide. 2. Hold the second slide on the top of slide with blood drop at 45° angle (as shown). 3. Hold the top slide and pull back to make contact with blood drop on the edge and let it spread by capillary action. 4. Push the top slide in forward direction in one motion while maintaining firm contact with bottom slide to produce the smear.

5. Clinical Significance

Complete blood count plays a major role in the identification of various diseases. A lower or higher than expected cell count may indicate the following.

a RBC

- ☆ Low RBC count indicates anemia which may be due to various factors including iron deficiency, external or internal haemorrhage, deficiency of Vitamin-12, bone marrow disorder and increased destruction of RBC in Spleen.
- ☆ High RBC count indicates Polycythemia- which is physiological in foetus, during exercise and at high altitude. Acute hypoxia increases RBC count and is caused due to smoking, asthma, blood disorder, dehydration, or because of steroids and/or diuretic drugs.

b. WBC

Types of Cells	*High Count may Indicate*	*Low Count may Indicate*
Neutrophils	**Neutrophilia:-** Bacterial infection, burns, stress, acute inflammation, injection of epinephrine, pyogenic infection, steroid treatment, chronic myeloid leukemia, splenectomy	**Neutropenia:-** Bone marrow disorder, chemotherapy, radiation, autoimmune disorder, infections like UTI, GIT, typhoid.
Eosinophils	**Eosinophilia**:- **Parasitic worm** infection, allergic reaction, ulcerative colitis, eczema, chronic infections, hematologicmalignancies (CML, Hodgkin's disease, eosinophilic leukemia) autoimmune disease like SLE.	**Eosinopenia:-** Alcohol intoxication, excessive production of cortisol (Cushing's syndrome), drug toxicity, corticosteroid injection, radiation, chemotherapy, bone marrow disorders
Basophils	**Basophilia**:- **Allergic reactions (food allergy, drug allergy, hay fever, allergic rhinitis)**, chronic myeloid leukemia, chronic inflammation, infections like chickenpox, tuberculosis.	**Basopenia**:- Hyperthyroidism, ovulation, pregnancy, radiation, chemotherapy, bone marrow disorders, steroid treatment.
Monocytes	**Monocytosis:- Chronic infection (viral or fungal**), Autoimmune disorder – Sarcodiasis, *etc.*	**Monocytopenia:-** Radiation, chemotherapy, bone marrow disorders, cortisol treatment.
Lymphocytes	**Lymphocytosis:- Acute viral infection,** chronic myeloid leukemia, acute infection- pertussis, protozoal infection (toxoplasmosis, trypanosomiasis),	**Lymphocytopenia**:- Prolong illness, Iimuno-suppression, HIV infection, steroid treatment, radiation, chemotherapy, bone marrow disorders

c. Platelets

Low Platelet count (Thromocytopenia)- Hypersplenism, Viral infections like dengue, HIV, Epstein Barr virus, chickenpox, leukemia, radiation, chemotherapy, bone marrow disorders, Vit-B12 deficiency, folate deficiency.

High platelet count (thrombocytosis)- Splenectomy, after trauma (surgery, injury, child birth), epinephrine injection *etc.*

6. Precautions

- ☆ Finger should be wiped properly before taking the blood.
- ☆ The angle of slide to form smear.
- ☆ The blood drop size should be proper.
- ☆ The pressure to form smear should be applied properly.
- ☆ Blood smear should be formed by single stroke.
- ☆ The smear should be air dried and fixed with absolute methanol prior to staining.

7. Questions

Multiple Choice Questions

1. **The condition for abnormally low levels of platelet count is known as**
   a. Thrombocythemia
   b. Leucopenia
   c. Thrombocytopenia
   d. Erythrocytopenia
2. **The blood cells including neutrophils, eosinophils and basophils are considered part of:**
   a. Agranulocytes
   b. Thrombocytes
   c. Granulocytes
   d. Megakaryocytes
3. **Which type of granulocytes kills the parasites and breaks the inflammatory substances?**
   a. Basophils
   b. Chlorophylls
   c. Eosinophils
   d. Neutrophils
4. **Which of the following cells consists of histamine in its granules present in cytoplasm**
   a. Neutrophils
   b. Basophils
   c. Eosinophils
   d. Monocytes
5. **Which of the following type of white blood cells kills parasitic worms and modulate allergic inflammatory responses?**
   a. Monocytes
   b. Lymphocytes
   c. Eosinophils
   d. Basophils
6. **Which of the following is not a function of neutrophils?**
   a. Scavenging cellular debris
   b. Mount immune response by direct cell attack or via antibodies
   c. Phagocytizes bacteria
   d. Undergo NETosis
7. **Which of the following is a function of basophils?**
   a. Release histamine during allergic reaction
   b. Killing of parasites
   c. Produce antibodies
   d. Phagocytosis of micro-organism
8. **Which of the following is performed by lymphocytes?**
   a. Maintains body temperature
   b. Secrete histamine
   c. Destruction of virus-containing cells
   d. Larvicidal activity

9. **Which of the following is the percentage of total volume used to determine the number of erythrocytes in a blood sample is referred to as the ________.**

   a. Erythrocyte sedimentation rate
   b. Red blood cell count
   c. Hematocrit
   d. Reticulocyte count

10. **The process by which neutrophils can produce reactive oxygen species to kill bacteria is known as ________.**

   a. Chemotaxis
   b. Amoeboid movement
   c. The respiratory burst
   d. Diapedesis

Short Answer Questions

1. **What is the normal percentage distribution of different types of white blood cells in blood circulation?**
2. **Define:**
   - ☆ **Diapedesis**
   - ☆ **Chemotaxis**
3. **What are the major pathological causes of leucocytosis and leucopenia?**
4. **Which leukocyte would be elevated in —**
   - ☆ **Bacterial infection**
   - ☆ **Viral infection**
5. **Describe the functions of lymphocytes.**

Long Answer Questions

1. **Briefly describe the structure and functions of different types of white blood cells.**
2. **What are the major causes associated with variations in different types of white blood cells?**

REFERENCES

Hall, J. E. (2011). Guyton and Hall Textbook of medical physiology (12th edition), Unit VI - Blood Cells, Immunity, and Blood Coagulation, Chapter 32-33, Saunders Elsevier; ISBN: 978-1-4160-4574-8. Pp 413-431.

Jacob, E. A. (2016). Complete Blood Cell Count and Peripheral Blood Film, Its Significant in Laboratory Medicine: A Review Study. American Journal of Laboratory Medicine, 1(3): 34-57. doi: 10.11648/j.ajlm.20160103.12.

Jiang, N., Tan, N. S., Bow Ho, Ding, J. L. (2007). Respiratory protein generated reactive oxygen species as antimicrobial strategy. Nat Immunol. 8(10): 1114-1122. doi: 10.1038/ni1501.

McMahon, T. J. (2019). Red Blood Cell Deformability, Vasoactive Mediators, and Adhesion. Front. Physiol. 10: 1-10. https: //doi.org/10.3389/fphys.2019.01417.

Prakash, G. (2012). Lab manual of blood analysis and medical diagnostics, S. Chand and Company Ltd., ISBN: 81-219-3967-4.

Sherwood, L. (2012). Introduction to Human physiology (8th Edition), Chapter 11- The blood, Brooks/Cole Cengage Learning; ISBN-13: 978-1-133-10454-4. Pp 409-435.Jain, A. K. (2003). Textbook of Physiology (7th Edition), Unit II- Blood, Avichal Publishing Company, ISBN: 81-7739-079-1. Pp 39-123.

4

Preparation of Single Cell Suspension of Splenocytes, Cell Counting and Viability Test

Dr. Sunita K. Yadav

ABSTRACT

Spleen is the largest secondary lymphoid organ. The cells present in the spleen are called splenocytes and they are a mixture of T and B lymphocytes. The splenocyte counting assay is performed to count the number of splenocytes during normal or abnormal conditions or after treatment of an animal with drugs or to assess the vaccine potential of an antigen. The splenocytes are counted using a hemocytometer and the dye (generally trypan blue) is used to assess the viability of the cells. The viable (live) cells exclude the dye after taking it while non-viable (dead) cells retain the dye, hence viable cells look colourless while non-viable cells look dark blue in colour.

Keywords: *Trypan blue, Cell suspension, Hemocytometer*

Learning Objectives

After the experiment, students will be able to:

- ✰ Identify and isolate the spleen.

- ✰ Prepare the splenocyte cell suspension
- ✰ Count the splenocytes using a hemocytometer
- ✰ Understand the importance of splenocyte counting assay

1. Introduction

The lymphoid organs are usually isolated to carry out experiments *in vitro* including the isolation of cells and subsequent culturing to observe the effect of various chemicals and environments. Various assays that can be carried out using lymphoid organ isolation are cell cytotoxicity assay, cytokine production analysis or response after chemical treatment, isolation of B or T lymphocytes (primarily present in lymphoid organs in large amount) *etc.* Among all lymphoid organs, the spleen is an important secondary lymphoid organ that helps in the activation and proliferation of immune cells (B and T lymphocytes) into effector and memory cells. It is an elongated and slightly curve-shaped organ that is located in the upper left abdominal quadrant. The convex ventral surface of spleen is smooth and dorsal surface is concave and attached to stomach by a gastro-splenic ligament. In a young adult mouse, the size of spleen is 15 mm (length) × 3 mm (width) × 2 mm (thickness) and weighs approximately ~100 mg. Around 5-15× 10^7 live lymphocytes can be isolated from spleen. The internal morphology of spleen has mainly two parts, red pulp, and white pulp, distributed in parenchyma (ground material). The red pulp is constituted by cords and vessels consisting of various stages of red blood cells. The young and flexible red blood cells pass through cords and continue to flow but older, larger, and deformed red blood cells are trapped by the splenic cords and undergo phagocytosis by macrophages present in the reticulum and sinus. The white pulp consists mainly of lymphoid tissues such as lymphocytes, plasma cells, macrophages, *etc.* The germinal centres in it are the sites for activated and proliferating lymphocytes.

2. Principle

Splenocytes consist of a population of T cells, B cells, dendritic cells and macrophages that are varied in their immunity function. It can be any white blood cell that is located or isolated from spleen. The viability of isolated splenocytes from spleen can be tested by using trypan blue (TB) exclusion assay. Trypan blue is a 0.9 kDa size chemical that is impermeable to the cell membrane and hence can enter only those cells that have degraded or compromised cell membrane. The entry and containment of trypan blue results in intracellular association and it imparted blue colour to the cell. If the cell is alive, its membrane inhibits the entry of trypan blue, and therefore alive cells looks whitish or non-bluish in color. The trypan blue exclusion assay works on the principle of excluding or binding trypan blue to live and dead cells respectively. The live cells remain unstained while dead cells look blue under a microscope. The counting of stained and unstained cells under a microscope gives the total no. of live and dead splenocytes for various assays. The RBCs from the spleen cell suspension are removed by using a lysis buffer containing ammonium chloride (NH_4Cl).

3. Requirements

- ☆ Biological materials: Mouse or rat
- ☆ Chemicals: 70 per cent alcohol, PBS buffer, lysis buffer (0.9 per cent ammonium chloride), 0.4 per cent trypan blue stain made in PBS buffer.
- ☆ Apparatus and tools: Dissection board, forceps, scissors, centrifuge tube, cell strainer, water bath, hemocytometer, cover slips, micropipettes, microscope (phase contrast preferably)

4. Procedure

4.1. Tissue (Spleen) Collection

1. Sacrifice the animal after injecting anaesthesia.
2. The animal should be kept on a clean dissection board and wiped with 70 per cent alcohol to remove any contamination on the surface.
3. Incise the abdominal cavity of the animal and remove the spleen, which is situated on in the left side of the abdomen, below the stomach.
4. Place the spleen into a tube (conical) containing chilled PBS buffer and label it properly.

4.2. Preparation of Splenocyte Cell Suspension (Figure 4a)

1. The excised spleen should be sliced in to small pieces.
2. Place the pieces between two slides and mash the pieces between them till they are converted in to single cell.
3. Place the cells in a tube and try to break the lumps by micropipette motions.
4. Centrifuge the cell suspension at 1500 rpm for 5 min to make the pellet.
5. Discard the supernatant.
6. The cell pellet should be resuspended in 2 ml of 0.9 per cent solution of NH_4Cl at 37 °C for 2 min to lyse the unwanted RBCs.
7. The cells are resuspended in PBS and PBS washes are given at 1500 rpm for 5 min.
8. Discard the supernatant after every wash and total 3 washes need to be given.
9. Finally, the cells are resuspended in 10 ml PBS for the counting assay.

4.3. Cell Viability Assay

1. Cell suspension (around 90 µl) should be put in a microcentrifuge tube using a serological pipette.
2. Add 10 µl of trypan blue and mix gently with the cells.
3. Clean the hemocytometer with 70 per cent ethanol and gently place the coverslip.
4. Draw some suspension containing trypan blue and release gently at the edge of the hemocytometer.
5. The 10× objective should be used to focus the grid lines of the hemocytometer, then focus on one set of 16 corner squares (shown in Figure 4b)
6. Count the no. of cells using a hand tally counter or manually in area of 16 squares.

   Note: Only count the live and healthy cells. The live cells are unstained. A rule of right-bottom (cells lying at right and bottom boundary line of the square) or left-upper (cells lying at left and upper boundary line) should be followed for all the squares while counting the cells. Don't apply both rules together. Count all 4 sets of 16 squares of hemocytometer in similar way.

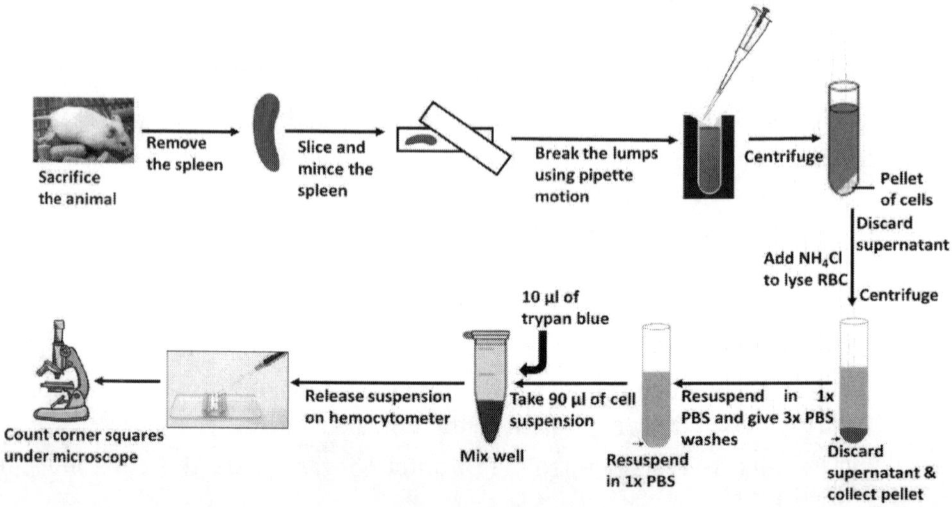

Figure 4a. Steps Showing Preparation of Splenocyte Suspension, Counting and Viability Test.

The no. of cells in 16 corner squares is equivalent to no. of cells counted $\times \mathbf{10^4}$/ **ml.**

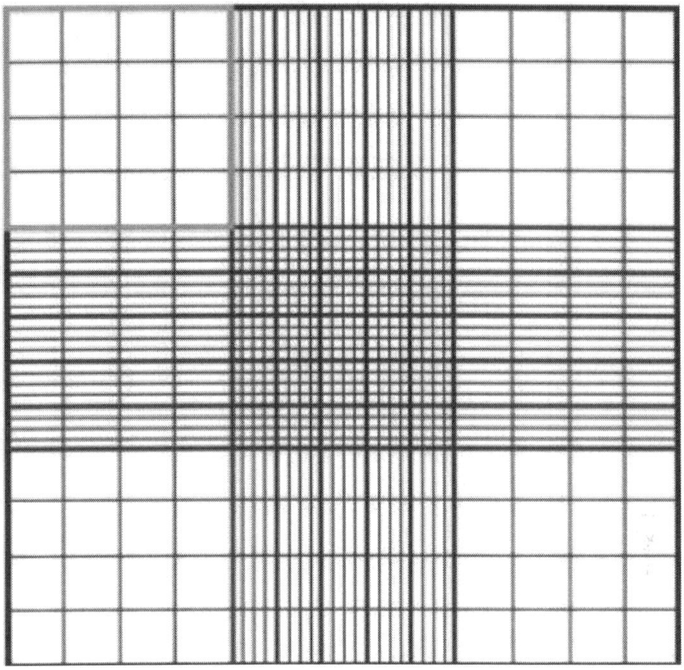

Figure 4b. Hemocytometer Gridlines.

The diagram of hemocytometer shows 16 corner squares in 4 sets that are used for cell counting.

Calculation

The formula for per cent cell viability is

$$\%\ \text{cell viability} = \frac{\text{Total viable cells (Unstained)}}{\text{Total cells (Viable+Dead)}} \times 100$$

$$\text{Viable cells per ml} = \frac{\text{Average of viable cell}}{\text{Square}} \times \text{dilution factor} \times 10^4\ \text{per ml}$$

$$\text{Dilution factor} = \frac{\text{Total volume (Volume of sample + Volume of diluting liquid)}}{\text{Volume of sample}}$$

Total live cell count:

- ☆ Count the total live cells in 16 corners of 4 sets.
- ☆ Divide the no. of cells by 4
- ☆ Multiply the count by dilution factor
- ☆ Multiply with the constant 10^4 to obtain cells/ml

- ☆ No. of live cells in 1st square: ....
- ☆ No. of live cells in 2nd square: ....
- ☆ No. of live cells in 3rd square: ....
- ☆ No. of live cells in 4th square: ....

Total number of live cells in all four squares: ......

Dilution factor : Final volume/initial volume

Initial volume : 90 µl

Trypan blue added : 10 µl

Therefore, final volume: initial volume + trypan blue = 90 µl+10 µl= 100 µl

$$\text{Hence, Dilution factor} = \frac{100\ \mu l}{90\ \mu l}$$

$$\therefore \text{Total number of viable cells} = \frac{\text{Total live cells}}{4} \times \frac{10}{9} \times 10^4 \text{ per ml}$$

Viability Assay

- ☆ The dead cells are stained by trypan blue so all the cells looking faint or dark blue within the grid should be counted and considered as dead cells.
- ☆ To check the viability:
  - Count live cells (unstained)
  - Count dead cells (stained blue with trypan blue)

$$\%\ \text{viability} = \frac{\text{Total live cell count}}{\text{Total cell count}} \times 100$$

5. Result

The total no. of live cells is found to be...... therefore the viability per cent is....

6. Applications

- ☆ Cell counting and viability assay help in accuracy, reproducibility and statistically significant results.
- ☆ It is also helpful in reviving and passaging of cells before setting the experiment related to cells
- ☆ It helps in determining cell health.
- ☆ Splenocyte counting and viability assay help in determining cell-mediated immunity.

7. Precautions

- ☆ Cell suspension should be made properly so that cells are not clumped and can be counted properly.
- ☆ Hemocytometer should not have bubbles while counting.
- ☆ After the cell counting, hemocytometer should be rinsed properly with distilled water or alcohol.
- ☆ The counting rule should be applied properly and each cell should be counted once.

8. Questions

Multiple Choice Questions

1. **The function of ammonium chloride while preparing cell suspension of splenocyte is:**
   - a) To remove proteins from the cell
   - b) To lyse unwanted RBCs
   - c) To increase the refractile index
   - d) To stabilise splenocytes
2. **The trypan blue dye**
   - a) Stains dead cells blue
   - b) Does not stain live cells
   - c) Both a and b
   - d) None of them
3. **Hemocytometer can be used to count:**
   - a) Any type of cells
   - b) Only splenocytes
   - c) Only RBC
   - d) Only WBC
4. **Which of the following is not a property of Trypan blue dye?**
   - a) Impermeable to cell membrane
   - b) Stains dead cells blue
   - c) Stain live cells red
   - d) Size is 0.9 kDa
5. **The splenocytes are counted in the_______of hemocytometer.**
   - a) All 4 corner squares containing 16 small squares in each corner square
   - b) The center square containing 25 small squares
   - c) Only one corner square is sufficient
   - d) Any of the above

Short Answer Questions

1. **Define**
   - ☆ **Cell viability**
   - ☆ **Cell suspension**
2. **Which part of hemocytometer is used for counting the cells?**

3. **What is the function of splenocytes?**
4. **What are the rules of counting using a hemocytometer?**
5. **Why NH_4Cl is added while preparing splenocytes cell suspension?**

Long Answer Questions

1. **What is the principle of cell counting?**
2. **Why the need arises to count the cells?**

REFERENCES

https: //www.abcam.com/protocols/counting-cells-using-a-haemocytometer

https: //www.colby.edu/bio/wp-content/uploads/sites/47/2014/01/Lab-3-splenocyte-prep-2013.pdf

https: //www.sigmaaldrich.com/IN/en/technical-documents/technical-article/cell-culture-and-cell-culture-analysis/mammalian-cell-culture/cell-quantification

https: //www.thermofisher.com/in/en/home/life-science/cell-analysis/cell-analysis-learning-center/immunology-at-work/immunology-protocols/mouse-spleen-cell-isolation-protocol.html

https: //www.surgery.pitt.edu/sites/default/files/GSVLLabInternal/Info_handbook/SOPs/General/Splenocyte per cent 20SOP.pdf

5

To Determine the ABO Grouping of the Blood Sample

Dr. Sunita K. Yadav

ABSTRACT

Type A, B, AB, and O phenotypes are the characterization of the ABO blood group system to determine the type of surface antigens and antibodies present in a donor's/acceptor's blood. The A and B are the types of antigens present on the RBCs surface. If the RBCs agglutinate, the matching antibody in the blood plasma must be missing. If antigens aren't found on RBC membranes, antibodies must be present in the blood plasma. Individuals older than 3-6 months have spontaneously existing antibodies that develop without any antigenic stimulation from the placenta of the mother. Other animals, such as gorillas and chimps also have the ABO blood group. For blood transfusion, compatibility of blood group is vital to save the lives of individuals, therefore, the determination of blood group is crucial for every individual.

***Keywords**: Antigen, Agglutination, Rh factor, Erythroblastosis foetalis, Donor.*

Learning Objectives

After the experiment, students will be able to:

- ☆ Perform the exercise to find their blood group
- ☆ Understand the principle behind blood group transfusion and importance of knowing the blood group

1. Introduction

During the transfusion of blood, the ABO blood group is an important parameter to assess the compatibility of blood groups between donor and recipient. The ABO blood group was discovered by Austrian scientist Karl Landsteiner in 1900. He identified O, A, and, B blood types and was awarded the Nobel prize in 1930 in physiology and medicine. The fourth type of ABO blood group, AB was discovered by Decastello and Adriano in 1902. All the individuals inherited one of the four blood groups *i.e.* A, AB, B and O and they have antibodies against the other blood group antigens. The antibodies, anti-A and anti-B antibodies are produced very early in life due to environmental substances sensitization. Few other animals such as rodents and apes also have an ABO blood type system. The epitopes in the ABO blood group system are due to sugars located on the red blood cell surface, N-acetylgalactosamine for the A-type and galactose for the B-type (Figure 5a).

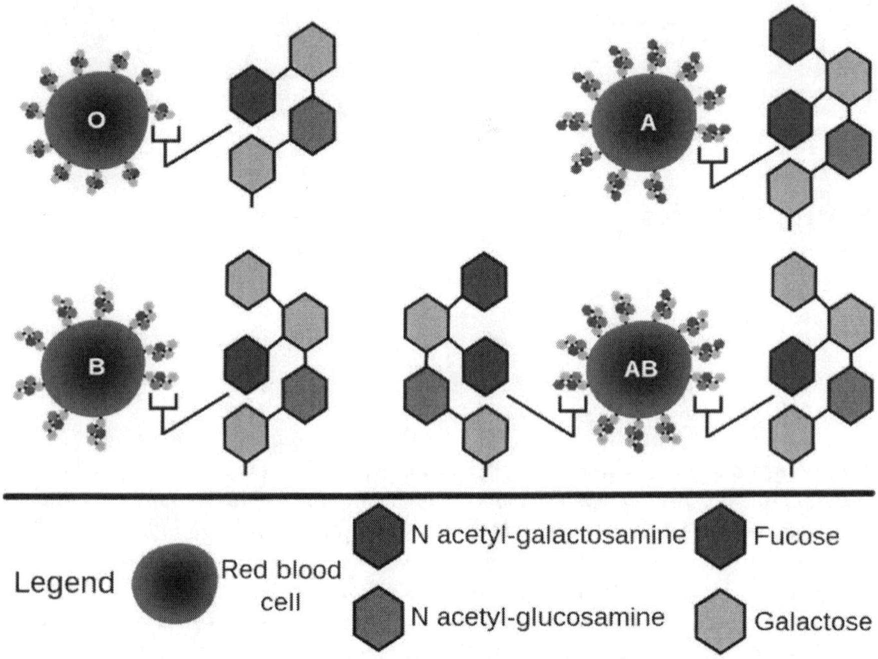

Figure 5a. Diagram Showing the Carbohydrate Chains that Determine the ABO Blood Group (*Source*: InvictaHOG, https://commons.wikimedia.org/w/index.php?curid=1161572).

The basis of the ABO blood group system is antigen-antibody interaction. Antigens are those molecules/particles that trigger antibody production. In the case of ABO blood grouping, red blood cells have exposed antigens on their surface in the form of carbohydrate moiety. Different antigens are present on RBC in the case of A and B blood group, both antigens in AB while no antigen in O blood group RBC. Therefore, people have immunological tolerance against their antigens present on their RBC, however, their immune system can produce antibodies against antigens present on different blood groups. These antibodies are called haemagglutinin and cause clumping of RBCs that carry foreign antigens.

The clumping and breaking down of RBC is a response shown by the adaptive immune system and is beneficial in case of infection. However, the severe reaction post blood transfusion due to different blood groups can result in the death of the recipient. The clumping occurs when A blood group individuals receive AB or B blood group, B blood group individuals receive A or AB blood group and O blood group people receive A, B or AB blood group. Since, A and B antigens on blood groups are modified forms of O blood group precursor, A, B and AB blood group individuals can receive blood from O blood group people (Figure 5b).

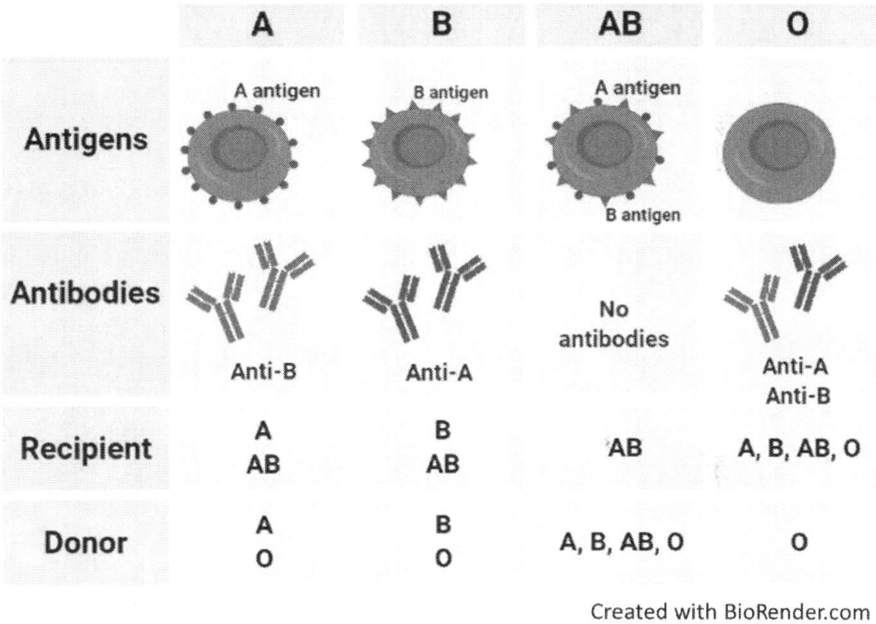

Figure 5b. Antigens and Antibodies Present in ABO Blood Group.

Another antigen, antigen D is also present on most people's red blood cells. It is also called as Rhesus factor or Rh factor and when it is present on RBC, the blood group is denoted as Rh positive while its absence from RBC surface makes the blood group Rh negative. An Rh- person does not have antibodies against Rh antigen but they can develop antibodies when their blood comes in contact with Rh+ blood group during blood transfusion. The Rh antigen present on the surface of Rh+ blood group triggers generation of anti-Rh antibodies in Rh- person plasma. Rh+ blood group people can receive blood from Rh- people but Rh- people could receive blood only from Rh- people only as they will develop an antigen-antibody reactions in their blood plasma if they receive blood from Rh+ people (Figure 5c).

S. No.	Your Blood Type	You can give blood to	You can receive blood from
1.	A+	A+ , AB+	A+ , A- , O+ , O-
2.	B+	B+ , AB+	B+ , B- , O+ , O-
3.	AB+	AB+	EVERYONE
4.	O+	O+ , A+ , B+ , AB+	O+ , O-
5.	A-	A+ , A- , AB+ , AB-	A- , O-
6.	B-	B+ , B- , AB+ , AB-	B- , O-
7.	AB-	AB+ , AB-	AB- , A- , B- , O-
8.	O-	EVERYONE	O-

Figure 5c. A Detailed Chart to Show Compatible Blood Groups during Transfusion.

2. Principle

The agglutination reaction is the basis for the ABO and Rh blood categorization systems. When red blood cells that carry one or both antigens (*i.e.* antigen A and antigen B) are exposed to the appropriate antibodies, they interact and generate observable agglutination or clumping. The antigens of the ABO blood groups are O-linked glycoproteins located on the surface of RBC exposing terminal sugar residue describe the type of antigen (A or B). Individual with blood group A has antigen-A on RBCs and anti-B antibodies in their serum. Similarly, people with blood type B have B antigen on RBCs and anti-A antibody in their blood. Individuals with blood group AB have both A and B antigens on their RBCs but no anti-A or anti-B antibodies in their blood while blood group O individuals possess both antibodies *i.e.* anti-A and anti-B antibodies in the serum but no antigen on RBCs surface. If a foreign antigen enters the body, the immune response of the person is triggered, and the WBCs are activated to identify the antigens. The WBCs then begin to produce antibodies against that antigen. Suppose, if A blood group person receives blood from a B group individual, the anti-B antibodies in the serum of A blood group individual bind and clump the antigen B residing on the surface of RBCs of the B blood group. The reaction results in agglutination of antigen B (means agglutination of RBCs) with anti-B antibodies leading to precipitation of the complex. The agglutination and hence precipitation reaction helps to identify the blood group of individuals.

3. Requirements

Glass-slides, Anti-A antibody, Anti-B antibody, Anti-D antibody (Figure 5d), Blood sample, Applicator sticks, Needle, Spirit lamp, 70 per cent alcohol, Tissue paper and cotton.

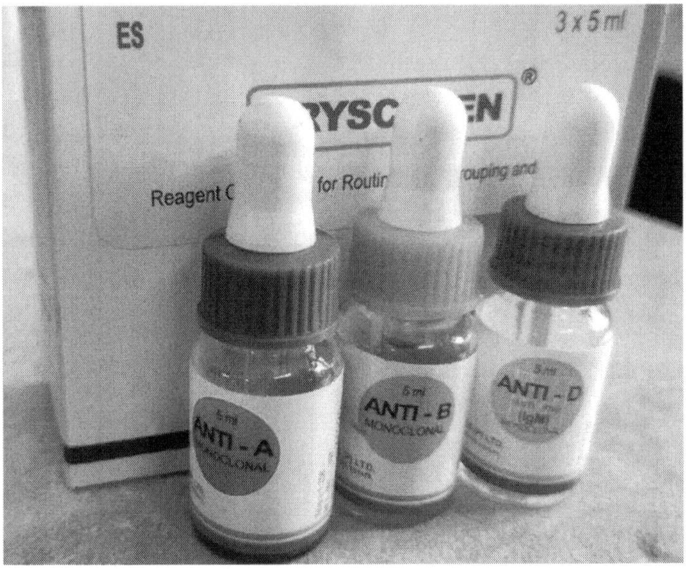

Figure 5d. A Kit Showing Anti-A, Anti-B and Anti-D Antibodies (Pic courtesy: Dr. Shailesh Kumar R. Yadav).

4. Procedure

1. Clean a glass slide and wipe it with 70 per cent alcohol
2. Wipe the finger with 70 per cent alcohol and prick it with a needle. Place a blood drop at three different areas of the glass slide (Figure 5e).
3. Name the areas on slides as A, B and D. Put anti-A antibody in area A, anti-B antibody in area B and anti-D antibody in area D.
4. Mix antibodies with a respective drop of blood with a wooden stick.
5. Leave the slide for a few seconds and observe the agglutination.

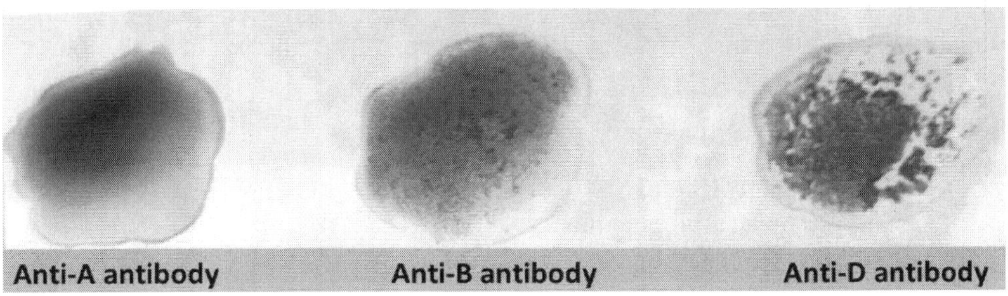

Figure 5e. Reaction of Antibodies to Blood Group. The slide shows B+ blood group reaction. (Pic courtesy: Dr. Shailesh Kumar R. Yadav).

5. Result

The blood group observed in the blood sample is __________

6. Discussion

The agglutination reaction is the result of an antigen-antibody reaction. The antibodies (anti-A, anti-B or anti-D) bind to the respective antigens present on the surface of RBC and result in cross-linking of many RBCs together. The antigen-antibody interaction results in the drop out of the complex from the solution. During a blood transfusion, the recipient may suffer a severe reaction if antibodies are present in his blood against donor's blood antigens. It may cause agglutination/hemolysis of donor's blood. However, blood of O-negative group people can be given to A, B and AB-blood group people (Rh positive and negative both) as the titer of antibodies in their blood (O-negative blood group people) against A and B-blood group of recipient are relatively lower and also they are diluted by the recipient's blood to a degree that they could not agglutinate/hemolyse the recipient's RBCs. Hence, O-negative people are called universal donors as they do not have any antigens on their RBC surface. Correspondingly, AB-positive blood group individuals are called universal acceptor (recipients) as they are devoid of any antibodies in their serum.

The anti-Rh antibodies need sensitization for their production in individuals. An Rh negative woman carrying a Rh positive fetus (inherited Rh trait from father) can become sensitized to Rh antigen while delivery of the baby or during miscarriage and abortion due to leakage across the placenta. The mother's immune system then produces anti-Rh antibodies of IgM isotype. Since sensitization occurs during fetus delivery, the first Rh positive child of a Rh negative mother does not show any complication. However, the successive Rh positive fetus suffers from hemolytic disease of newborn wherein anti-Rh antibodies can cross the placenta and attack fetus RBCs leading to high bilirubin pigment in the blood of fetus resulting in jaundice. The disease is also called erythroblastosis fetalis. It can be prevented by an injection named Rhogam (anti-Rh antibodies) that is given to Rh negative mothers within 3 days (72 h) of delivery. The antibodies bind the circulating RBCs of fetus and destroy them before they could sensitize the mother's immune system against Rh antigen.

7. Applications

- ☆ Blood group test is carried out to determine the compatibility of the donor and acceptor's blood group during transfusion process.
- ☆ It is also used to solve paternity dispute cases.
- ☆ Used to detect infant hemolytic diseases.
- ☆ Predisposition to a variety of disorders such as peptic ulcer generally belongs to the O blood group category people.

8. Precautions

- ☆ The fingers and slides should be wiped with 70 per cent alcohol to prevent any contamination of blood or slide while doing the experiment.
- ☆ Blood and antibodies should be mixed properly and observed carefully to avoid any mistake in the interpretation of result.
- ☆ The antibodies used in the experiment should be stored properly in the fridge to increase the shelf life.
- ☆ The needle used by students to prick the finger should not be mixed and reused.
- ☆ Blood-slides should not be cleaned off at the sink. They should be placed into a 10 per cent solution of Clorox before discarding.
- ☆ The antibodies should not be mixed with each other.
- ☆ The blood grouping reaction should be carried out at 20-30 °C.

9. Questions

Multiple Choice Questions

1. During ABO blood grouping test, agglutination is visible with anti-A and anti-D antibody, which blood group is it?

a. A+ b. A-

c. B+ d. B-

2. The agglutination reaction depends on:

a. Antigen-antigen interaction

b. Hemoglobin- antibody interaction

c. Antigen-antibody interaction

d. Both a and c

3. In ABO blood grouping, the blood group O is due to:

a. Presence of antigen O

b. Presence of antigen A and antigen B

c. Absence of antigen A and antigen B

d. Presence of antigen O, antigen A and antigen B

4. Which of the following is false about ABO blood grouping?

a. Blood group O people do not have antigen A and antigen B on RBCs

b. Blood group type AB people have both antibody A and antibody B in their plasma

c. Blood group type B people have antigen B on their RBCs

d. Blood group type AB+ individuals are universal acceptor

5. **During blood transfusion process, if the blood group of person receiving blood is unknown, then which blood group can be donated to him?**

   a. AB+ b. AB-

   c. O+ d. O-

Short Answer Questions

1. **If mother is A+ and father is B+, what are the probable blood groups that their progeny can inherit?**
2. **Why ABO blood grouping test is required during blood transfusion?**
3. **Who discovered ABO blood grouping?**
4. **Define:**
   - ☆ **Rh factor**
   - ☆ **Rhogam**
   - ☆ **Hemolysis**
   - ☆ **Agglutination**
5. **Which antigens and antibodies are present in individuals having following blood groups?**
   - ☆ **AB+**
   - ☆ **A+**
   - ☆ **O-**
6. **Which blood group is considered a universal donor?**
7. **Which blood group individuals can accept blood group from any person?**

Long Answer Questions

1. **Explain the principle of ABO blood grouping.**
2. **Give applications of the ABO blood grouping test.**
3. **What is erythroblastosis foetalis? How it can be prevented?**

REFERENCES

https: //himedialabs.com/TD/HTI008.pdf

https: //laboratorytests.org/reverse-blood-grouping/

http: //nbtc.naco.gov.in/assets/resources/training/5.pdf

Raman, L.; Armstrong, B.; Smart, E. Principles of laboratory techniques. *ISBT Sci. Ser.* **2008**, *3*, 33–60

6

Detection of Complement Activity Using Hemolysis of Antibody Coated SRBC and Standard Serum

Dr. Sunita K. Yadav

ABSTRACT

The complement system is a collection of proteins that functions in immune system response. Activation of these proteins of complement system aid in the lysis of target cell and assist in phagocytosis by opsonisation. The complement activity is a screening assay to assess the activation of the classical complement pathway. Reduction, absence and non-functioning of complement proteins affect the complement activity assay. It examines the potential of complement proteins of the classical pathway present in the serum to lyse sheep red blood cells (SRBC) pre-coated with anti-SRBC antibody raised in other species (rabbit). The antibody coated SRBC incubation with serum (needs to be tested) leads to activation of classical pathway of complement proteins and causes lysis of SRBC. As lysis causes release of haemoglobin in the solution, it can be used to measure the activity of complement proteins in the serum. Therefore, if complement proteins are absent, the complement activity will be zero and if there is a reduction in components of complement pathway, the absorbance of haemoglobin will be lower indicating less complement activity.

Keywords: *Complement proteins, Opsonisation, Phagocytosis, Hemoglobin.*

Learning Objectives

After the experiment, students will be able to:

- ✰ Understand the immune reaction aided by complement proteins
- ✰ Perform complement assay in the laboratory
- ✰ Understand the importance of complement proteins in activation of immune system

1. Introduction

The complement system is a collection of proteins that, when activated, cause target cell lysis and aid phagocytosis via opsonization. The complement system consists of around 30 proteins located in blood plasma or on the cell surface. They are components of innate immune system that protect an organism from foreign particles or pathogens. In the 19th century, it was found that human serum free from cells could kill pathogenic microbes. Pathogens, when enter in their host organism are killed by lysis and the lytic activity of blood serum, can be enhanced by immunization with the same pathogen prior to its natural entry. The heated serum (at 56°C for 30 min.) could agglutinate the pathogen but cannot kill it, the lytic activity can be restored if fresh serum is mixed with heated serum. Even fresh serum taken from naive animal can restore the lytic activity back. Therefore, the lytic activity depends on two factors: a heat-stable factor that is present in immunized animal and a heat-labile factor that is present in both, naive and immunized animals. The heat-stable factor was identified as antibody molecules that are generated in immunized animals (adaptive immunity) while the heat-labile factor was recognized as complement proteins, present in naïve and immunized animals (innate immunity). The three pathways adopted by complement system to clear the pathogen are classical, alternative and lectin pathway. The alternative and lectin pathways function without adaptive immunity involvement while classical pathway requires specific antibodies for functioning. Therefore, classical pathway act as a mediator between innate and adaptive immunity.

Antibodies that have adhered to the surface of an invading cell activate the classical route. The membrane-bound antibody triggers the activation of the first component of complement pathway, which in turn triggers the activation of eight other complement proteins. The end outcome is the development of the membrane attack complex, a collection of proteins that creates a pore in the membrane, allowing target cells to be lysed.

Functions of complement proteins:

- ✰ Coating of antigen-antibody complex (opsonization) to augment the phagocytosis process.
- ✰ Help in the lysis of bacteria and cells that are infected
- ✰ Regulates antibody response and clears immune complexes
- ✰ Can cause inflammation and anaphylaxis reaction.

2. Principle

The experiment described here demonstrates and quantifies the activity of serum complement proteins in a simple manner. RBCs in vertebrates contain substantial levels of the respiratory pigment in the cytoplasm called haemoglobin, which is immediately identified when the cell released it upon lysis. The experimental diluent turns red after RBC lysis, and the intensity of the red color, which corresponds to the amount of haemoglobin liberated, may be detected spectrophotometrically. Briefly, antibody is taken from a rabbit that has been heavily vaccinated with cell membranes (stroma) of sheep RBCs (SRBCs). The antibody used for the experiment should not have any other component of serum (complement proteins should be heat inactivated). The antibody solution is subsequently treated with whole SRBCs in the absence of complement proteins. Complement binds to the antibody's highly conserved constant domain, allowing immunoglobulins from various species to bind and activate complement proteins. The sensitized (antibody-coated) erythrocytes are next treated with guinea pig serum, which serves as a source of complement in this experiment. Hemoglobin is released into the buffer once the SRBCs are lysed by activated complement. Target SRBCs that were not lysed by complement are centrifuged and the intensity of the red color in the supernatant is measured at 540 nm. The amount of haemoglobin released by complement-mediated hemolysis is proportional to the observed absorbance (rupture of RBCs). The positive control is the total lysis of RBCs with ammonium chloride and the negative control is RBCs incubated with dilution buffer.

The total complement activity/CH50 level will be zero if a complement component is missing; the CH50 level will be reduced if one or more components of the classical pathway are reduced. Each serum dilution contains a predetermined volume of optimally sensitized SRBC. The ability of various dilutions of test serum to lyse antibody coated SRBC is used to measure the quantity of complement activity.

3. Requirements

- ☆ Barbitone-buffered saline (containing calcium and magnesium ions)
- ☆ Sheep erythrocytes (SRBC) in Alsever's solution
- ☆ Ammonia solution (0.04 per cent)
- ☆ Fresh serum or preserved guinea-pig serum
- ☆ Horse serum for anti-erythrocyte antibody
- ☆ Equipment: Spectrophotometer

4. Procedure

4.1. Standardization and Sensitization of Erythrocytes

1. Wash 4 ml of sheep erythrocyte suspension by spinning (generally supplied in 25 per cent v/v in Alsever's solution) in barbitone buffered saline for 3 min at 200 × g.

2. Resuspend the washed erythrocytes in barbitone-buffered saline (15 ml).
3. One ml of erythrocytes is lysed by mixing with 25 ml of ammonia solution. Read the absorbance of solution at 540 nm (absorbance after lysis should be 0.48-0.50 in 1 cm cuvette for 6 per cent SRBC suspension; so, suspension can be adjusted accordingly).
4. Mix 15 ml of 6 per cent SRBC with equal volume (15 ml) of barbitone-buffered saline and 0.1 ml of horse hemolytic serum and incubate the mixture at 37°C for 15 min.

   Note: Use the mixture within 24 h.

4.2. Estimation of Complement Activity

1. Set the tubes as shown in Table 7. In the positive control tube (not shown in table), add 1.20 ml ammonium chloride solution to 0.3 ml sensitized erythrocytes suspension.
2. Use fresh serum as a complement source.
3. Place the tube on ice and 2 ml of buffer should be added to each tube.
4. Centrifuge the tube at 200 g for 10 min at 4°C.
5. The supernatant (1 ml) from each tube is removed and absorbance should be taken at 540 nm.

Table 7. Test tube set-up

Tube No.	*Buffer (ml)*	*Guinea-pig Serum (ml) (initial dil. 1:30)*	*Sensitized Erythrocyte Suspension (ml)*
1	1.20	—	0.3 (negative control)
2	1.10	0.10	0.3
3	1.05	0.15	0.3
4	1.00	0.20	0.3
5	0.90	0.30	0.3
6	0.80	0.40	0.3

Source: Practical Immunology by Hay *et al.*, 2002. Blackwell Science Ltd.

4.3. Calculation of Results

1. Calculate the percentage lysis in each tube by taking positive control as 100 per cent lysis.
2. Plot percentage lysis on y-axis against complement concentration (ml of undiluted serum) on x-axis. A sigmoid curve will be obtained as shown in Figure 6
3. For each complement dilution (serum dilution), plot log x against log[y/(100-y)]. The straight line thus obtained has a slope of 1/n. The abscissa

intercept of the line (where log[y/(100-y)] =0, is the log dilution resulting in 50 per cent lysis.

Note: Complement components are heat labile so fresh serum should be used for activity assay.

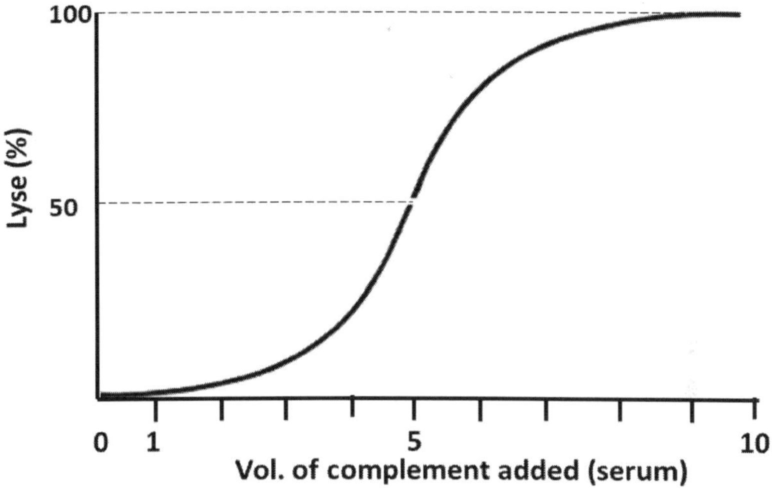

Figure 6. Sigmoid Curve Showing Lysis of Sheep Erythrocytes (SRBC) Sensitized by Horse Anti SRBC in the Presence of Complement Proteins (*Source*: Practical Immunology by Hay *et al.*, 2002. Blackwell Science Ltd).

5. Results

The per cent lysis in the experimental tube was found to be....

The CH50 value from the graph was found to be..........

6. Discussion

The complement assay in a patient's serum is most often done to diagnose autoimmune disorders such as rheumatoid arthritis, lupus and sometimes to diagnose bacterial, viral and fungal infections. The lower amounts of complement proteins with a decreased activity means the person is suffering from autoimmune disorders or a pathogenic infection or some other ailment. In certain types of cancers such as lymphoma, the activity and amount of complement proteins can be increased. However, the complement assay has various sources of error. Sometimes, spontaneous hemolysis occurs leading to false results. Also, during the sensitization step, different source of antibody has different affinity to SRBC resulting in dissimilar results. The collected specimen for complement activity should be protected from heat as many complement proteins are heat labile and high temperature leads to their inactivation.

7. Applications

Complement assay is generally used

- To diagnose or monitor autoimmune disorders such as Lupus and Rheumatoid arthritis
- To diagnose viral, bacterial or fungal infections
- To detect certain types of cancers such as leukemia or lymphoma in which complement activity is increased.
- Evaluate the effect of drug on inhibition or usage of complement proteins.

8. Precautions

- Blood should be taken out carefully.
- All the steps need to be done at proper temperature.
- The sample mix should be used within 24 h.

9. Questions

Multiple Choice Questions

1. **In the complement pathway, cell lysis is initiated by:**
   a) Formation of membrane lysis complex
   b) Formation of membrane destructive complex
   c) Formation of membrane damaging complex
   d) Formation of membrane attacking complex
2. **Which of the following is required for complement activation?**
   a) Ca b) Mn
   c) Zn d) Cl
3. **The determination of complement activity assay depends on**
   a) Activation of RBCs b) Lysis of WBCs
   c) Activation of WBCs d) Lysis of RBCs
4. **The positive control in complement activity assay is:**
   a) Total lysis of antibodies b) Total lysis of hemoglobin
   c) Total lysis of RBCs d) Total lysis of WBCs
5. **In the complement activity assay explained here, the horse serum is the source of:**
   a) Complement proteins b) RBCs
   c) Antigens d) Anti-SRBC antibody

Short Answer Questions

1. **Define**
   - ☆ **Complement proteins**
   - ☆ **Total complement activity**
   - ☆ **Sensitized SRBC**
2. **In which medical condition, complement activity is increased?**
3. **When a person is suffering from decreased complement activity?**
4. **At which wavelength, the absorbance is recorded in the complement activity assay?**

Long Answer Questions

1. **Explain the principle of complement activity assay.**
2. **Give applications of complement activity assay.**

REFERENCES

Costabile M. Measuring the 50 per cent haemolytic complement (CH50) activity of serum. *J Vis Exp.* 2010;(37): 1923. Published 2010 Mar 29. doi: 10.3791/1923

https: //www.sinobiological.com/research/complement-system/total-complement-activity

Jordan E. Inglis, Kimberly A. Radziwon, and Gregory D. Maniero. The serum complement system: a simplified laboratory exercise to measure the activity of an important component of the immune system. Advances in Physiology Education 2008 32: 4, 317-321

Practical Immunology Frank C. Hay, Olwyn M.R. Westwood and Paul N. Nelson © 1976, 1980, 1989, 2002 by Blackwell Science Ltd

7

Enzyme-Linked Immunosorbent Assay (ELISA)

Dr. Sunita K. Yadav

ABSTRACT

The enzyme-linked immunosorbent assay (also known as EIA: Enzyme Immunoassay) is an important assay technique for detecting and measuring antibodies, peptides, proteins, and hormones in the blood. It is based on the principle of protein-protein interaction. Antibodies present in the serum are created by B lymphocytes in response to a specific antigen. In the event of some infections, it aids in determining the existence of specific antibodies in the body to detect the infectious agent. ELISA differs from other antibody assays in that it provides qualitative as well as quantitative findings.

Keywords: *Antigen, Antibody, 96-well plate, Blocking, Absorbance.*

Learning Objectives

After the ELISA experiment, students would be able to:

1. Perform the procedure of carrying out ELISA for various detection assays
2. Differentiate between types of ELISA
3. Understand the importance of the ELISA method in clinical settings

1. Introduction

The enzyme-linked immunosorbent assay (ELISA) is an immunological technique that is used to measure the concentration of antibodies or antigens present in a sample. In the past, during 1960s when ELISA technique was not known, immunoassay was usually conducted using radioactively labelled antibodies or antigens and hence called radioimmunoassay. The signal provided by radioactivity labelled antigen/antibody helps in detection of their presence in the solution. Few years after, an alternative to radioactivity (hazardous to health) was discovered in the form of enzymes that can be ligated to antigen or antibody to provide a signal in the form of colour. The knowledge of enzyme ligated antibody/antigen led to the development of a method named ELISA by Peter Perlman and Eva Engvall at Stockholm University of Sweden in 1971 and by another group, the Dutch research group of Anton Schuurs and Bauke Van Weemen at the same time. A lot of progress has been achieved since then and newer ELISA techniques use fluorogenic and chemiluminiscent types reporter systems to quantify analytes (proteins/antigens/antibodies) in the sample that are more sensitive as compared to traditional ELISA. The general principle of these techniques is similar to ELISA, therefore despite the absence of ligated enzyme to analyte, they are also called ELISA.

2. Principle

ELISA is a biochemistry assay used to detect the presence of specific analytes (antigens/antibodies) in a liquid solution. Analytes are those substances whose qualitative or quantitative presence is being detected in a given sample. The assay involves the addition of samples and chemicals in a sequential manner resulting in a signal that is quantifiable as a measure of specific analyte present in the solution. ELISA works on the principle of antigen-antibody reaction on an immobilized solid surface that could be detected by the enzyme ligated to the detection reagent added during the process resulting in colour change. The process involves addition of multiple reagents and washing of any unbound analyte/reagent at each step. Those analytes/reagents that are specific, remain bound while non-specific reagents/analytes are washed away leading to specific quantitative estimation of the unknown sample. The quantitative estimation involves the measurement of colour intensity (produced by enzymatic reaction ligated to the detection reagent) at a particular wavelength by spectrophotometry. In ELISA, as the enzymes are employed for detection, thus, it was named as Enzyme-linked immunosorbent assay. Since the analytes are also called ligands as they could specifically bind to detection reagent, therefore ELISA is included in the ligand binding assays category. In ELISA, usually antibody titer is measured that are raised against a specific antigen in bulk quantity or specific antigens can be detected by using specific antibodies such as in various disease diagnoses (COVID 19, Dengue virus detection *etc.*)

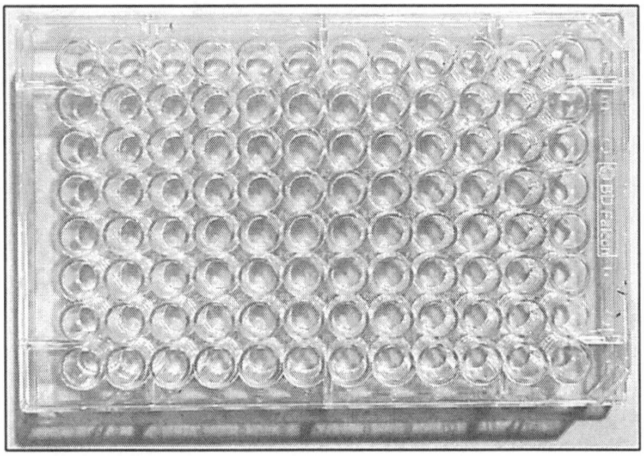

Figure 7a. A Multiwell (96-well microtiter plate) ELISA Plate (Pic courtesy: Dr. Shailesh Kumar R. Yadav).

2.1. Types of ELISA

In the ELISA experiment, the first step is the adsorbance of antigen/antibody present in the solution or sample to the surface of ELISA plate (microtiter plate) (Figure 7a). The formation of the antigen-antibody complex is detected by detection antibody that is directly conjugated to enzyme such as alkaline phosphatase (AP). Sometimes the antibody (called primary antibody) directly binding to antigen is labelled with enzyme or it is allowed to bind to the secondary antibody called detection antibody. Based on the sequence of antigen or antibody added during ELISA procedure, it can be divided in to the following 4 categories (Figure 7b):

2.1.1. Direct ELISA

An antigen is immobilized on the surface of microtiter plate in the case of direct ELISA. The immobilized antigen is then detected directly by antibody conjugated to an enzyme such as AP. The sequence of molecules can also be reversed wherein antibody is coated to the microtiter plate and an antigen labelled with enzyme is used for detection. Since only few steps are required in direct ELISA, it is faster as compared to other ELISA techniques. This type of ELISA is used for measuring an immune response to an antigen.

2.1.2. Indirect ELISA

In indirect ELISA, antigen is adsorbed to a well in microtiter plate. The first step is binding of unlabeled antibody (primary antibody) to the specific antigen coated to wells followed by detection using detection antibody (secondary antibody) conjugated to an enzyme. The detection antibody is specific to the primary antibody and required for detection purpose only. This method is best for estimating total antibody concentrations in the blood.

2.1.3. Sandwich ELISA

As the name suggests, an antigen is sandwiched between primary antibody (capture antibody) and secondary antibody (detection antibody). Both antibodies are specific to antigen but to non-overlapping regions or epitopes of the antigen. Therefore, antibody pairs should be selected carefully to ensure that they bind to different regions of antigen. The secondary antibody has an enzyme conjugated to it in most cases, if not then a detection antibody is provided ligated to an enzyme.

2.1.4. Direct and Indirect Sandwich ELISA

In this process, the first step is to coat the capture antibody to the surface of wells of ELISA plate followed by adding analyte/antigen (sample) to the well. To detect the signal, detection antibody is added which is conjugated with enzyme. Such cases are called direct sandwich ELISA. Sometimes, the detection antibody is unlabeled and hence a third antibody conjugated with enzyme is required. This type of sandwich ELISA is called indirect sandwich ELISA.

Sandwich ELISA is generally used to measure the antigen quantity in complex samples where crude antigens are present in place of purified. For *e.g.* elevated/ depressed cytokine levels can be measured in an immune response.

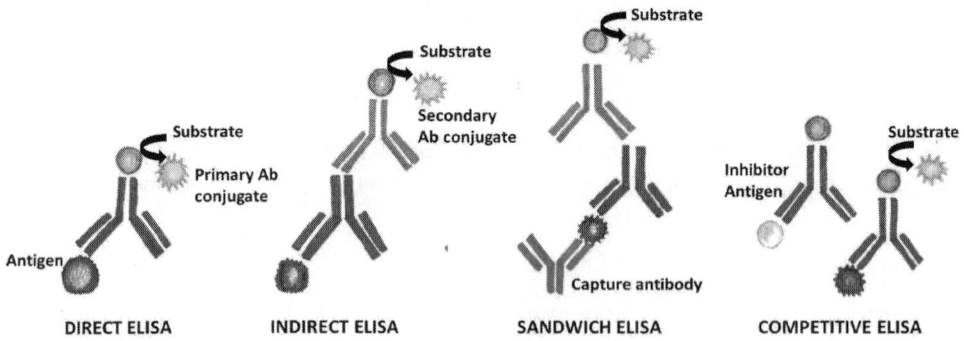

Figure 7b. Different Types of ELISA
(Illustrator: Aakriti Singh, Department of Botany, University of Delhi).

2.1.5. Competitive ELISA or Inhibition ELISA

This ELISA is the most complex of all ELISA assays. The competitive/inhibition ELISA is mainly used to estimate the concentration of antigen/antibody in a sample by measuring interference in the output signal. The analyte that is to be measured competes with a reference antigen or antibody to bind to the limited labelled antibody or antigen respectively. The higher the concentration of analyte in a sample, the more of these analytes compete with reference antigen or antibody and hence bind to the labelled antibody or antigen respectively. The observed reduction in signal gives an idea of high amount of analyte in the tested sample. The output signal negatively correlates with the amount of analyte present.

It is generally used when only one antibody is present for a single antigen or in case of small antigens that are incapable to bind to two different antibodies.

Table 8. Advantages and Disadvantages of different Types of ELISA

Sl.No.	*Type of ELISA*	*Advantages*	*Disadvantages*
1.	**Direct ELISA**	As the technique has fewer steps, it is faster than other ELISA	Higher background due to non-specific binding of all proteins to the plate
		Since fewer reagents and steps are involved, it is less prone to error	In the absence of any signal amplification, it is less sensitive
			Since each protein required a specific antibody, it has less flexibility.
2.	**Indirect ELISA**	Since more than one labelled detection antibody can bind to primary antibody, the assay is highly sensitive	High background noise in case if detection antibody is cross reactive.
		Inexpensive assay as fewer labelled antibodies are required	Since secondary antibody is needed, it is a longer procedure as compared to Direct ELISA
		Since different primary antibodies can bind to single labelled detection antibody, it shows greater flexibility	
3.	**Sandwich ELISA**	It has high sensitivity as compared to Direct/indirect ELISA	Since two or more than two antibodies are involved, the cross reactivity occurs sometimes therefore, the assay needs standardization.
		Since two or more than two antibodies are involved in detection, it has a high specificity	
		This assay is flexible as both direct and indirect detection can be done.	
4.	**Competitive ELISA or Inhibition ELISA**	No need of purified sample. Crude and impure sample can be used easily.	It cannot be used in case of a diluted sample or the sample is less as it has low specificity
		Since little deviation from ELISA assay has no or very less effect on the outcome, therefore it is more robust as compared to other types of ELISA.	
		Since the variability is nil or negligible in duplicate assays, it has high consistency.	
		It is highly flexible ELISA as any type of ELISA assay can be modified according to it.	

3. Requirements

- ☆ Multiwell plate (96 wells) (Figure 7a)
- ☆ Antigen (Protein), Antibodies (Primary antibody, Secondary antibody conjugated with enzyme)
- ☆ 1×PBS, 1×PBST buffer
- ☆ 3 per cent BSA
- ☆ DD water
- ☆ ELISA plate reader, micropipettes, tips, measuring cylinder, flask or beaker
- ☆ Substrate for enzyme

4. Procedure

Since there are several types of ELISA, the procedure varies but the underlying principle is the same for all. All types of ELISA have already been explained in the principle section and the procedure is described using a well labelled diagram. Here, a basic procedure of indirect ELISA is explained which is the most popular method to carry out ELISA (Figure 7c)

1. The analyte present in the sample is coated on the microtiter plate in the concentration of 100 µg/ml/well made in 1 × PBS for O/N at 4 °C. (analyte will bind to the ELISA plate via passive adsorption)
2. Next day, wash the plate with 1× PBST three times for 10 min at RT.
3. A blocking solution, generally casein or 3 per cent BSA made in 1× PBST is added to each well for 2 h at 37 °C to cover non-specific sites and to reduce the background. Give three washes with 1×PBST after incubation period is over.
4. Then add primary antibody made in 1×PBS against analyte to each well at the required dilution and incubate the plate for 1 h at 37 °C.
5. After 1 h, wash the plate with 1× PBST thrice for 10 min each at RT.
6. Add detection (secondary antibody) antibody made in 1× PBS at required concentration (or according to the manufacturer's protocol) and incubate for 1 h at 37 °C. Wash the plate with 1 × PBST thrice after an hour.
7. Add specific substrate of the detection antibody (according to manufacturer) and observe for any colour change for 5-30 min. Stop the reaction using acid (H_2SO_4 or H_3PO_4).
8. Measure the absorbance at an optimum wavelength in ELISA plate reader at 450 nm (in case if detection antibody is conjugated with AP).

   (*Note*: The detection antibody is ligated with an enzyme that is usually alkaline phosphatase (AP) or horseradish peroxidise (HRP). These enzymes carry out the reaction by catalysing the hydrolysis of a chromogenic substrate, such as p-Nitrophenyl Phosphate, Disodium salt

(PNPP) or 3,3',5,5'-tetramethylbenzidine (TMB) respectively and convert them in to product that is colourful and could be measured using an ELISA (spectrophotometric) plate reader at specific wavelength.

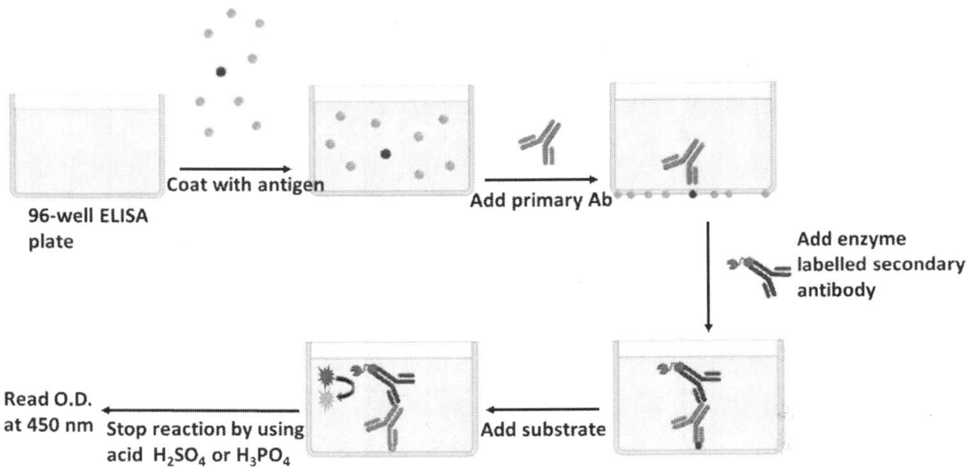

Figure 7c. Steps of Typical ELISA Assay (*Source*: Created by Biorender.com).

5. Discussion

The ELISA assay output can be of three types based on the results.

1. Qualitative: The presence or absence of an antigen/antibody can be detected by ELISA when a blank or control well has no or unrelated antigen/antibody.
2. Quantitative: The concentration of an antigen or antibody can be detected by ELISA by plotting a standard curve in the form of serial dilution of known concentration of the analyte and then comparing the result. ELISA assay can be employed for accurate estimation of the unknown concentration of an analyte.
3. Semi-quantitative: The relative levels of antigen can be measured in samples by ELISA assay wherein the colour signal is directly proportional to the concentration of the antigen.

The sigmoidal curve to estimate concentration is plotted on a graph taking optical density Vs log concentration. The standard curve is plotted by considering known concentrations of antigen and unknown samples are measured by comparing linear portion of the curve. Such calculations can be done manually or with the help of software available in ELISA plate readers.

6. Sensitivity of ELISA Assay

Among all immunoassays, ELISA is the most sensitive detection assay. The range to measure molecules accurately lies between 0.01 ng to 0.1 ng. The result depends upon the efficient antigen-antibody interactions. If the signals are weak, better chemiluminescent or signal enhancing fluroscent dyes or molecules can be employed for better results.

7. Applications of ELISA

ELISA can be used to detect:

- ✰ Infectious viral diseases by measuring antigens present in them
- ✰ Levels of hormones present in body fluid.
- ✰ Various types of bacterial or parasitic infections.
- ✰ Presence of drugs in the body
- ✰ Presence of allergens in several types of food such as milk, eggs, peanuts *etc.*
- ✰ Adulterants and toxins present in food
- ✰ Presence of antibodies in serum

8. Precautions

- ✰ PBST washes should be given properly at each step.
- ✰ Blocking should be done to remove non-specific binding
- ✰ Antibody dilution should be optimum
- ✰ The enzymatic reaction at the last step should be stopped with a stop reaction.
- ✰ Pipetting error should be avoided. Use multichannel pipettes for PBST wash and to add the reagents quickly.

9. Questions

Multiple Choice Questions

1. The goat anti-rabbit IgG horseradish peroxidase conjugate in the ELISA experiment acts as:

a) Primary antibody b) Substrate

c) Blocking reagent d) Secondary antibody

2. The product color in ELISA reaction is ______to the concentration of antigen in the sample.

a) Inversely proportional b) Directly proportional

c) Independent d) Depends on substrate

3. Which of the following cannot be detected by ELISA?

a) Detection of HIV antibodies in patient's blood

b) Detection of Hepatitis B infection

c) Detection of *mycobacterium* antibodies

d) Hemoglobin percentage in blood

4. Which step is followed after blocking step in ELISA?

a) Substrate addition

b) Detection

c) Addition of antibody

d) Coating the plate

5. The incubation of 96-well ELISA plate with antigen or antibody at the start of experiment is called:

a) Sandwiching

b) Detection

c) Coating

d) Blocking

Short Answer Questions

1. **Expand ELISA.**
2. **Differentiate between**
   - ☆ **Direct ELISA and Indirect ELISA.**
   - ☆ **Sandwich ELISA and Competitive ELISA.**
3. **Why blocking is required in ELISA procedure?**
4. **Give the sensitivity range of ELISA.**
5. **Name two enzymes that are generally conjugated with secondary antibody (detection antibody) in ELISA.**

Long Answer Questions

1. **What is the principle of ELISA?**
2. **Explain different types of ELISA.**
3. **What is the use of ELISA?**

REFERENCES

Alhajj M, Farhana A. Enzyme Linked Immunosorbent Assay. [Updated 2022 Feb 2]. In: StatPearls [Internet]. Treasure Island (FL): StatPearls Publishing; 2022 Jan-. Available from: https: //www.ncbi.nlm.nih.gov/books/NBK555922/

https: //assets.thermofisher.com/TFS-Assets/LSG/Application-Notes/TR0065-ELISA-guide.pdf

https: //www.abcam.com/kits/elisa-principle

Kindt T, Goldsby R, Osborne B. 2007. *Kuby immunology.* W.H. Freeman and Co., New York, NY

Raman, L.; Armstrong, B.; Smart, E. Principles of laboratory techniques. *ISBT Sci. Ser.* 2008, *3*, 33–60

8

To Learn the Technique and Basic Patterns of Precipitation by Ouchterlony's Double Immuno-diffusion Method

Dr. Sunita K. Yadav

ABSTRACT

Immuno-diffusion is a technique that employs diffusion through a medium such as agar or gel agarose for the detection or quantification of antibodies and antigens. It is one of the numerous methods for producing a precipitate between an antibody and its antigen of interest. Among various types of immunodiffusion, Ouchterlony double diffusion is carried out on gel where antibodies when react with antigen, result in formation of antigen-antibody complexes that can be seen as a line of precipitation. Since it is a double immunodiffusion, both the antigen and antibody are allowed to diffuse into the gel. The method is applicable to detect various pathogens in serological testing as well as to analyse the specificity of the antibody to an antigen.

Keywords*: Antigen-antibody interaction, Precipitation, Precipitin arc.*

Learning Objectives

After the experiment, students would be able to:

- ☆ Detect antigen-antibody complexes.
- ☆ Explain the principle behind antigen-antibody complex formation and precipitation.
- ☆ Recognize antigen-specific antibody
- ☆ Assess the antigen similarity

1. Introduction

A variety of techniques are based on gel immunodiffusion to analyse the antigen-antibody interaction. Antigen-antibody forms a complex when they react to each other. The complex formation depends on various parameters such as type of antigen/antibody, concentration, specificity and the ratio of them in the reaction.

Among numerous diffusion techniques, Ouchterlony double diffusion (ODD) is a simple technique that is used to assess the presence of a specific antibody against an antigen in the collected antisera and to determine titer of the antibody. Hence, the chief use of this assay is in qualitative diagnostic procedures to detect specific microorganisms by identifying its antigen with the help of specific antibody. The ODD was first developed by Dr. Orjan Ouchterlony in 1948. The "double diffusion" name has come from the procedure of the technique where both antigen-antibody are allowed to diffuse/migrate towards each other in a gel. The place where the two reactants (antigen and antibody) meet, a line is formed due to precipitation. The precipitin that is formed is highly specific and only forms when a specific antibody reacts with its antigen. The advantage of the Ouchterlony double diffusion method is that the experiment can be carried out with limited resources but the main drawback is that the technique is less sensitive as compared to other diagnostic methods.

The antigens are biomolecules that are usually multivalent (numerous antigenic determinants) to which many antibodies can bind at the same time. Since each antibody has minimum of two variable regions to bind antigen, large aggregates are formed of antigen-antibody resulting in precipitation. The occurrence of cross-linking and hence precipitation depends on the optimum amount of antigen and antibody present in the solution. The precipitation will not occur if any of the two is in excess. Therefore, it is desirable that an increasing amount of antigen in a stepwise manner should be added to a constant amount of antibody in solution. The excess amount of antibody is referred as antibody excess zone (called prozone). The concentrations of antigen and antibody are highest at the well where they are loaded, however, when they start diffusing, the concentration keeps decreasing. The point where they meet each other, antigen will react to their specific antibody to form an Ag-Ab complex. At the zone of equivalence, the concentration of antigen and antibody are in optimal ratio and the precipitated complex is more. Excess of antigen leads to decrease in the formation of precipitation, such zone is called antigen excess zone. Therefore, the amount of antibody and antigen should be in optimum concentration for complex formation that will lead to precipitation.

The specificity of the antibody to antigen decides the various geometrical patterns that are produced between antigen and antibody wells. The patterns of lines are useful to assess the difference between antigens. The various types of patterns are explained below (Figure 8a):

Pattern A: Fusion of Lines at the Junction to Form a Proper Arc

The antibodies in the antisera react with both the antigens causing precipitation along a smooth line. In this case, both antigens are similar or immunologically identical hence the antibodies reacted equally with both of them.

Pattern B: Crossed Lines

In the case of antigens that are non-identical/non-similar, none of the antibodies cross-react with antigenic determinants that might present on both antigens. Hence, the two antigens are unrelated immunologically according to the antibodies' reaction.

Pattern C: Fusion of Two Lines with a Spur

In the case of antigens that are showing partial identity with each other, the antibodies in the antiserum cross react more with one antigen as compared to another. The determinants are assumed to form spur in one antigen while other antigens could not form spur due to lack of those determinants.

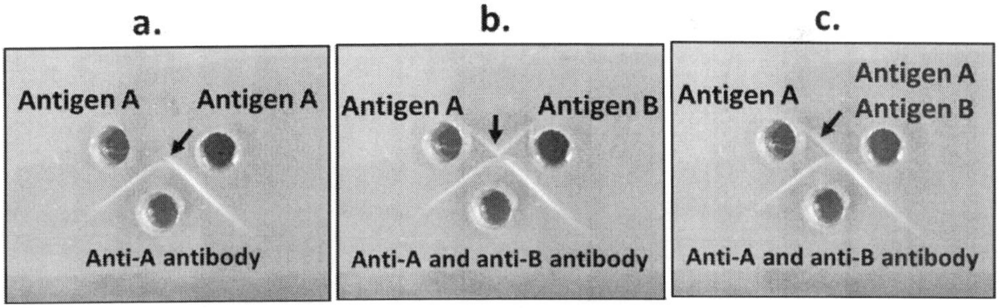

Figure 8a. Various Patterns of Precipitation.

Panel a. A case of identical antigens (Identity). Panel b. A case of non-identical antigens (Non-identity). Panel c. A case of partial identical antigens (partial identity) (Pic courtesy: Dr. Shailesh Kumar R. Yadav).

2. Principle

The Ouchterlony double diffusion technique is based on the principle of antigen-antibody interaction. In this procedure, both *i.e.* antigen and antibody loaded in the wells are allowed to diffuse towards each other, and hence, it is called double diffusion. When antigen and antibody migrate towards each other, a zone of equivalence is achieved where the concentration of antigen and antibody is optimum to react with each other to form an immune complex resulting in precipitation. The line of precipitation is formed when antigen and antibody are complementary to each other and the pattern of the line gives an idea about the antigen similarity and antibody titer. Since antigens are generally multivalent (several antigenic determinants to which antibodies bind), precipitation reaction is favoured. All the antibodies have minimum two antigen binding sites (IgM antibody has 10); therefore, large complexes of antigen-antibody are formed that result in precipitation due to their huge size. When antigen and antibody are in optimal concentration, their reaction leads to high precipitation and hence the zone formed around the area is called zone of equivalence (Figure 8b). The amount of precipitation is lowest in case of excess antibody (prozone) and excess antigen (antigen excess zone)

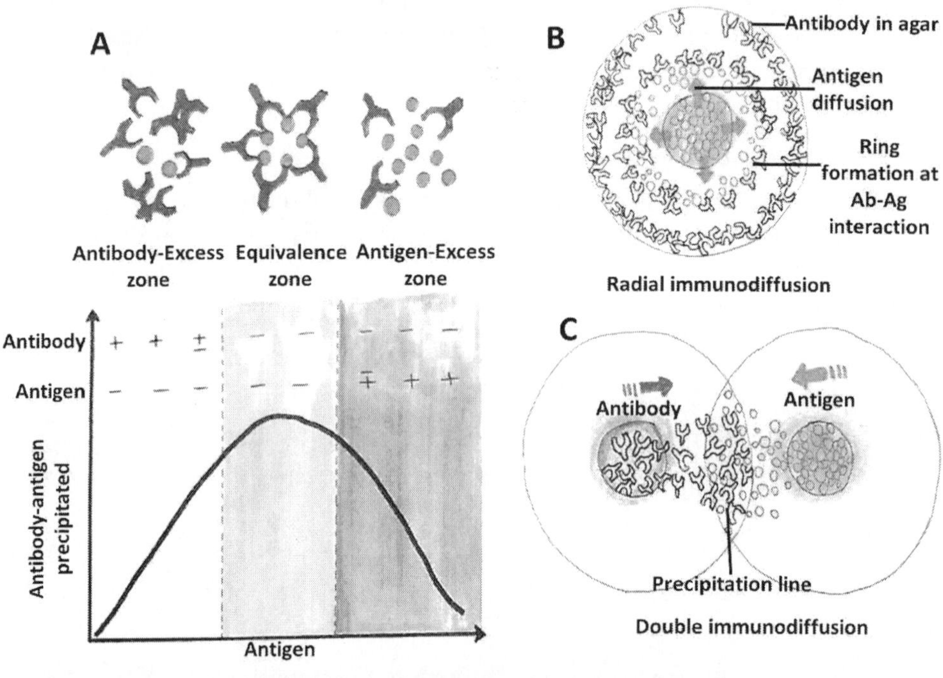

Figure 8b. (A) Precipitation curve showing various antibody and antigen zones. (B) Radial immunodiffusion (C) Double immunodiffusion (Illustrator: Aakriti Singh, Department of Botany, University of Delhi).

3. Requirements

Agarose (low EEO), 1×PBS (pH 7.4), glass plate, gel punch with syringe, antisera, antigens, template, saline, distilled water, conical flask, measuring cylinder, micropipettes, tips.

4. Procedure

1. An agarose gel of 1.2 per cent (~30 ml) should be prepared in 1× PBS buffer. Dissolve agarose powder thoroughly by boiling in microwave oven or on hot plate.
2. Pour slightly cooled solution (50-60 °C) on glass plates (4 ml/plate) placed on smooth horizontal surface. Let the gel set for 30 min.
3. Keep the glass plate on template and punch the wells of appropriate size.
4. Fill the wells with antiserum and antigen (10 µl each) as shown in Figure 8a.
5. Place the glass plates in a moist environment overnight at 37 °C.
6. Observe the result.

5. Result

The incubation overnight resulted in the formation of opaque lines showing antibody-antigen precipitation due to complex formation. Various patterns are obtained that will help in identification of identical/similar or different antigens.

6 Discussion

The pattern of precipitin line formed by interaction of antigen-antibody helps in determining the specificity of antibody to antigen. If a single continuous line is formed, it indicates the homogeneity of antigens to the antibody. In case of heterogeneity of antigens to the antibody, more than one precipitin line is formed that shows the presence of different epitopes (non-identical) on the antigens.

7. Applications

- It is used to detect, identify and quantitate the antigens and antibodies.
- It is used for determining the similarity between two different antigens and hence, a tool that can be used in evolutionary studies.
- It is used for determining the antibody titer.
- It is used in serological testing of various pathogenic organisms.
- It is also employed for identifying fungal antigens

8. Precautions

- A warm (not hot) agarose should be poured.
- No air bubble should be trapped in-between agarose.
- Agarose should be allowed to solidify without disturbance before starting the experiment.
- Wells should be labelled properly to avoid confusion.

- ✰ Glass-plate should be placed overnight undisturbed and covered in moist environment.
- ✰ Use separate tips for different reagents.

9. Questions

Multiple Choice Questions

1. **The pattern of partial identity is observed in Ouchterlony double diffusion method when:**
   a. Two antigens are identical
   b. Two antigens are similar
   c. Two antigens are unrelated
   d. None of the above
2. **Ouchterlony double diffusion technique does not require:**
   a. Red blood cells
   b. Antibodies
   c. Antigen
   d. An agar plate
3. **Which antibody shows the best agglutination reaction?**
   a. IgG
   b. IgD
   c. IgE
   d. IgM
4. **Precipitation between antigen-antibody occurs at:**
   a. Prozone
   b. Zone of equivalence
   c. Antigen excess zone
   d. Antibody excess zone
5. **Antigenic determinants are:**
   a. The sites on antigen where antibodies bind
   b. The sites on antibody where antigens bind
   c. Proteins that define size of antigen
   d. Proteins that define size of antibody

Short Answer Questions

1. **What is the meaning of crossed lines in the set-up of Ouchterlony double diffusion method?**
2. **Why Ouchterlony double diffusion method is named so?**
3. **What happens in the equivalence zone in Ouchterlony double diffusion method?**

Long Answer Questions

1. **What is the principle of Ouchterlony double diffusion method?**
2. **Give applications of Ouchterlony double diffusion method.**

REFERENCES

https: //www.creative-diagnostics.com/Immunoassay.htm

https: //himedialabs.com/TD/HTI002.pdf

https: //webstor.srmist.edu.in/web_assets/downloads/2021/18BTC106J-lab-manual.pdf

Kindt T, Goldsby R, Osborne B. 2007. *Kuby immunology.* W.H. Freeman and Co., New York, NY

Mujtaba MG, Baliban T, Bhagu J, Herrera M. 2021. A laboratory exercise simulating antibody and antigen reactions of the Ouchterlony double immunodiffusion assay using inorganic salts. J Microbiol Biol Educ 22: e00103-21. https: //doi.org/10.1128/jmbe.00103-21.

Practical Immunology Frank C. Hay, Olwyn M.R. Westwood and Paul N. Nelson © 1976, 1980, 1989, 2002 by Blackwell Science Ltd

9

To Learn the Technique of Immuno-electrophoresis

Dr. Sunita K. Yadav

ABSTRACT

Immunoelectrophoresis is the term used to describe the precipitation of proteins in agar under the influence of an electric field. The term "immunoelectrophoresis" was coined by Grabar and Williams in 1953. This process combines immuno-diffusion and electrophoresis technique. Electrophoresis is used to separate an antigen combination into its constituent elements, and then double immuno-diffusion is used to test the mixture. A sample (mix of antigens/proteins) is electrophoresed on gel and molecules are separated based on their charges. The gel is then sliced into a trough, and antibodies are loaded in it. Antibodies in the gel diffuse laterally to meet diffusing antigens/proteins, forming complexes and resulted in precipitation, allowing for antigen identification.

Keywords: *Electrophoresis, Trough, Cathode, Diffusion, Precipitation*

Learning Objectives

After the experiment, students would be able to:

- ☆ Understand the principle of Immunoelectrophoresis technique

- ☆ Perform the Immunoelectrophoresis assay in laboratory
- ☆ Differentiate among different types of Immunoelectrophoresis.

1. Introduction

Immunoelectrophoresis is a qualitative technique to characterize different types of antibodies. The sample containing antigen mixture is separated by agarose gel electrophoresis based on the charge. The monoclonal antibody or serum is then allowed to diffuse laterally within the gel. The specific antibodies bind to respective antigens on the gel to form a white precipitation arc. These arc are difficult to see and a black background is used to see them.

The serum levels of important immunoglobulins *i.e.* IgM, IgG and IgA can be determined by using immunoelectrophoresis technique. Since the technique comprises separation as well as diffusion of antigens towards their respective antibodies and vice-versa, it is used majorly to determine the occurrence, specificity, homogeneity and detection of high level of immunoglobulins.

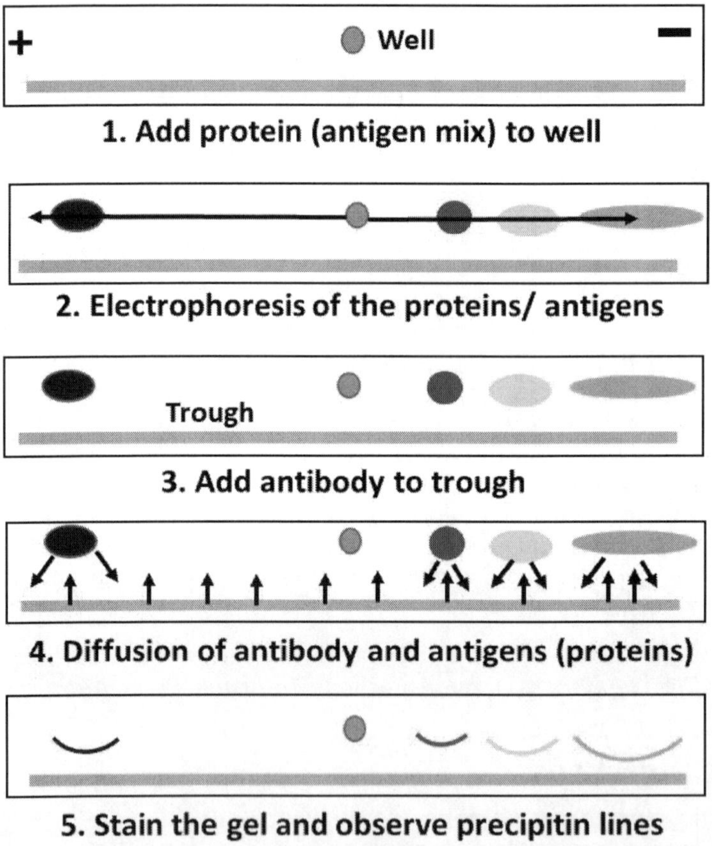

Figure 9a. Immunoelectrophoresis Technique.

2. Principle

It involves the reaction of antigen to its corresponding antibody resulting in the development of precipitate lines at the point where equivalence zone of antigen and antibody reach. The sample containing antigen is separated in to its constituents by electrophoresis on the basis of their charge. The antisera added in the troughs diffuse in the gel to react with antigen. Since the antigen and antibody diffuse in the gel, the point where ratio of antigen-antibody is optimum, a complex is formed by the reaction of antigen to its corresponding antibody causing the precipitation of a white complex in the gel that looks like an arc (Figure 9a).

Immunoelectrophoresis can be categorized in to following three types (Figure 9b):

a. **Counter immunoelectrophoresis**: This type of immunoelectrophoresis as explained by Estela and Heinrichs involve counter reaction of antigen and antibody where antibody moves by electroosmotic flow as no charge is present on them at pH 8.6 (provided by buffer) in agarose gel. However, the charged antigens move towards antibodies by electrophoresis resulting in the formation of white precipitate on agarose gel.

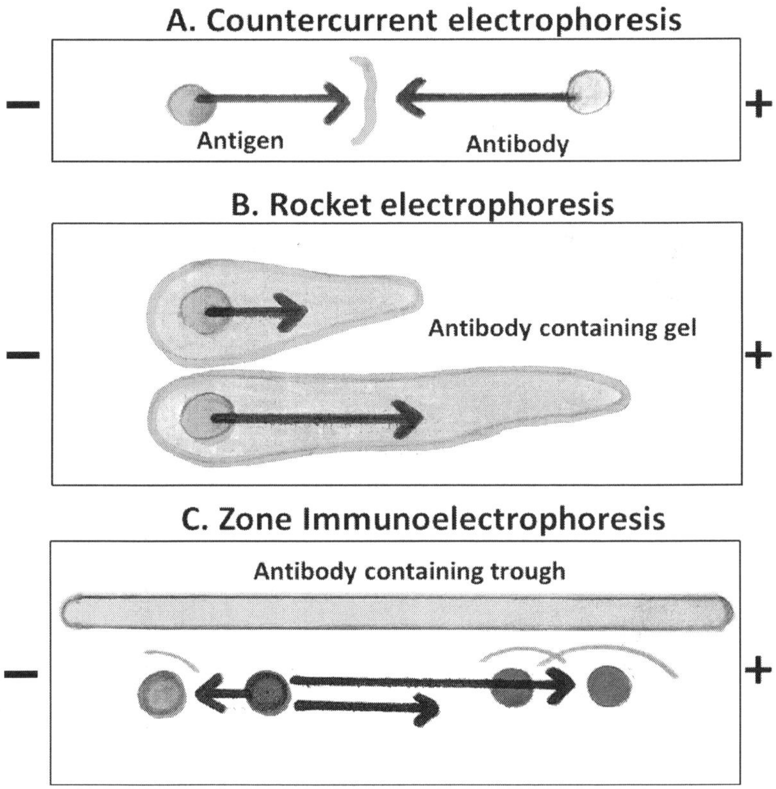

Figure 9b. Different Types of Immunoelectrophoresis.

b. **Zone electrophoresis immunodiffusion**: Graber and William explained that in this type of electrophoresis, antigens are separated on agarose gel by electrophoresis. Then they are allowed to diffuse towards antibodies that are filled in the trough in the parallel lane.

c. **Rocket electrophoresis immunodiffusion**: This type of immuno-electrophoresis was explained by Laurell. A definite concentration of antibody is placed in an agarose gel towards which antigens migrate in the agarose gel. In this method also, antibodies are not charged due to buffer. Antibody and antigen migrate towards each other and keep forming an immunocomplex until an equivalence zone is reached. The result is that rocket shaped precipitin lines are formed and the enclosed areas are proportional to the concentration of antigen.

3. Materials

Conical flask, measuring cylinder, distilled water, micropipettes, micropipette tips, gel cutter, casting tray, agarose, gel punch, Agarose (EEO-high), antigen, antiserum A, antiserum B, electrophoresis apparatus, template, electrophoresis buffer (1×)

Buffer Preparation

Sl.No.	*Chemicals*	*Quantity*
1.	Diethyl barbituryl acid	3 gm
2.	Calcium lactate	3 gm
3.	Diethyl sodium barbiturate	15.5 gm
4.	Sodium azide	1 gm

Note: Dissolve Diethyl barbituryl acid in hot double distilled water. Then add Diethyl sodium barbiturate. When it dissolves completely, add calcium lactate and sodium azide and make up the volume till 1 liter mark. Adjust pH to 8.6.

4. Procedure

4.1. Agarose Gel Slide Preparation

1. Prepare 1 per cent agarose gel (0.5 gm agarose in 50 ml of 1× electrophoresis buffer, heat till agarose melts completely).
2. Premark the slide towards the negative electrode (cathode)
3. Pour the cooled agarose solution on casting tray placed on a horizontal flat surface. Avoid any bubble formation.

   (*Note*: In case of uneven or problematic pouring, recollect, reheat and repour the agarose solution)
4. Place the tray on a template to make punches. Punch the well using a gel puncher towards anode.

5. Use a gel cutter to make two troughs.

   (*Note*: Do not remove the gel from the trough)

4.2. Electrophoresis

1. Seal the wells with 4-5 µl of molten agarose and allow it to cool
2. Add 25µl antigen to the well. Place the casting tray in electrophoresis chamber containing 1×electrophoresis buffer.

   (*Note*: Antigen well should be at the cathode side)
3. Run the gel for 45 min at 50 volts.

4.3. Immunodiffusion Process

1. Afterward electrophoresis, remove the gel from each trough and place the slide at room temperature for 15 min. Seal the troughs with a minimum volume of molten agarose.
2. Add 50 µl of antiserum A and antiserum B in separate troughs
3. Allow the diffusion of antigen and antiserum by placing the casting tray in a humidified chamber at room temperature for 24 h.

5. Results

The formation of an elliptical precipitin arc shows an interaction between antigen and antiserum while no precipitation suggests no reaction between antigen-antiserum.

6. Discussion

The antigen and antibody react to each other to form a complex that precipitates in the solution, hence the precipitin line suggests that the antibody present in antiserum can recognise specific antigen and bind to it. The intensity, shape and positioning of precipitin line are useful to identify different antigens or proteins. The heterogeneity of antigens can also be identified using the immunoelectrophoresis method.

7. Applications

- ☆ Can be used to detect immunodeficiency.
- ☆ Excess production of serum proteins or less secretion of complement proteins
- ☆ Free particles of monoclonal antibodies
- ☆ To identify and detect changes in serum or urine proteins
- ☆ Useful to monitor the purity of antigens and help in identification of a single antigen from a mixture of antigens samples.

Advantages

- Since immunoelectrophoresis combines electrophoresis for the separation of antigens and immunodiffusion for reaction between antigen and antibody, it is an analytical technique with great resolving power.
- It is useful to identify several antigens in the serum at same time.

Limitations

- Immunoelectrophoresis technique is slower and less sensitive compared to immunofixation electrophoresis. The results of immunoelectrophoresis are also difficult to interpret sometimes.
- It has limited use in food analysis.

8. Precautions

- A warm (not hot) agarose should be poured.
- No air bubble should be trapped in-between agarose.
- Agarose should be allowed to solidify without disturbance before starting the experiment.
- Tips should be changed frequently to prevent the mixing of reagents.

9 Questions

Multiple Choice Questions

1. In counter-current electrophoresis technique:

a. Electrophoresis help in the movement of antigen and antibody toward each other

b. Electrophoresis help in migration of antibody toward negative charge

c. Electrophoresis help in driving away antigen

d. Electrophoresis help in the movement of antigen and antibody parallel to each other

2. Immunoelectrophoresis is a combination of:

1. Precipitation and agglutination
2. Immunological assay and SDS-gel electrophoresis
3. Immunoprecipitation and agar diffusion
4. Immunodiffusion and electrophoresis

3. In immunodiffusion technique, ____ is added to the well and ____ is added to the troughs of the agar gel.

a. Serum and antigen b. Antibody and antiserum

c. Antiserum and serum d. Serum and antiserum

4. **The antigen mixture during electrophoresis is separated based on their:**
   a. Molecular mass
   b. Charge
   c. Size
   d. All of them
5. **The diffusion reaction between antigen and antibody in immuno-electrophoresis is allowed for:**
   a. 2-4 h
   b. 4-8 h
   c. 8-16 h
   d. 18-24 h

Short Answer Questions

1. **Define immunoelectrophoresis**
2. **In immunoelectrophoresis, what is added to the wells and what is added to the troughs?**
3. **Which buffer is used for immunoelectrophoresis?**
4. **How long the plates are incubated and in what type of environment?**
5. **The wells made in immunoelectrophoresis are towards which electrical charge?**

Long Answer Questions

1. **What are the different methods on which location of bands depends?**
2. **What are the advantages of immunoelectrophoresis?**

References

Academia.edu

Csako G. Immunoelectrophoresis: A Method with Many Faces. Methods Mol Biol. 2019;1855: 249-268. doi: 10.1007/978-1-4939-8793-1_21. PMID: 30426422.

https: //webstor.srmist.edu.in/web_assets/downloads/2021/18BTC106J-lab-manual.pdf

Practical Immunology Frank C. Hay, Olwyn M.R. Westwood and Paul N. Nelson © 1976, 1980, 1989, 2002 by Blackwell Science Ltd.

Glossary

Adaptive immunity: Immunity that has memory and occurs after exposure to an antigen either from a pathogen or a vaccination

Agar: It (also called agar-agar) is a jelly-like substance consisting of polysaccharides obtained from the cell walls of some species of red algae. Among various applications, it is used to solidify the medium.

Agarose: It is a heteropolysaccharide, generally extracted from certain red seaweed. It is a linear polymer made up of the repeating unit of agarobiose, which is a disaccharide made up of D-galactose and 3,6-anhydro-L-galactopyranose. It is used to separate DNA and RNA during electrophoresis.

Agglutination: An antigen-antibody reaction in which antigen combines with its antibody in optimum conditions resulting in the formation of visible clumping particles.

Amplicon: It is a piece of DNA or RNA that is the source and/or product of amplification or replication events.

Analytes: The chemical substance measured and/or identified in a test sample.

Anaphylaxis: An acute allergic reaction to an antigen to which the body has become hypersensitive.

Annealing: It is heteroduplex formation from two complementary (or nearly complementary) molecules or regions of single-stranded DNA (ssDNA)

Antibodies: A blood protein produced in response to an antigen and counteract it.

Antigen: A toxic or other foreign substance that induces an immune response in the body, especially the production of antibodies.

Antigenic determinants: An antigenic determinant is the site of an antigen molecule that is termed an epitope and interacts with the specific antigen-binding site in the variable region of an antibody molecule known as a paratope.

Aqueous phase: The homogeneous part of a heterogeneous system that consists of water or a solution in water of a substance. In genomic DNA/plasmid DNA isolation, the upper part after chloroform: isoamyl alcohol addition, is referred to as an aqueous phase.

Blotting: It is a technique by which a macromolecule such as DNA, RNA, or protein is resolved in a gel matrix, transferred to a solid support, and detected with a specific probe.

Cell suspension: Suspension culture is a type of culture in which single cells or small aggregates of cells multiply while suspended in an agitated liquid medium

Chemotaxis: In terms of microbiology, it refers to the migration of cells toward attractant chemicals or away from repellents.

Clumping: Formation of lumps or compact mass, occurs when antibodies combine with RBC.

Colony forming unit (CFU, cfu or Cfu): It is a unit that estimates the number of microbial cells (bacteria, fungi, viruses etc.) in a sample that are viable and able to multiply via binary fission under controlled conditions.

Competence: The ability of bacteria to take up foreign DNA

Complement system: It is a system of plasma proteins that can be activated directly by pathogens or indirectly by pathogen-bound antibody.

Conjugation: It is the process by which one bacterium transfers genetic material to another through direct contact.

Depurination: It is one of the most prevalent forms of DNA damage in which the N-glycosidic bonds are cleaved to release the corresponding adenine or guanine from DNA.

Diapedesis: It is the process in which white blood cells come out of the blood vessels into the surrounding area in case of injuries.

Dilution factor: It (also known as the dilution ratio) is the ratio between the final volume and initial volume of the solution.

DNA fingerprinting: It (also called DNA typing, DNA profiling, genetic fingerprinting, genotyping, or identity testing) is a method of isolating and identifying variable elements within the base-pair sequence of DNA (deoxyribonucleic acid) to determine an individual's DNA characteristics.

Double Immunodiffusion: In this method, both the antigen and antibody diffuse independently through agar gel in two dimensions, horizontally and vertically.

Electrophoresis: It is a laboratory technique used to separate DNA, RNA or protein molecules based on their size and electrical charge.

Electroporation: A method used to introduce DNA into cells using high-voltage electric shocks.

Epitope: It is the part of an antigen molecule to which an antibody attaches itself.

Equivalence zone: It is the zone where the concentration of antigen and antibody are in optimal ratio and the precipitated complex is highest.

Erythroblastosis foetalis: It is a severe medical condition that results from an incompatibility between certain blood types of pregnant woman (Rh-) and her foetus (Rh+) destryoing foetal RBCs.

Harvest: It is the process of collecting target cells from a growth medium. It is usually carried out through centrifugation.

Hemolysis: It is the breakdown of red blood cells.

Hybridization: In term of genomics, it is the process in which two complementary single-stranded DNA and/or RNA molecules bond together to form a double-stranded molecule.

Immunoelectrophoresis: It is a process of a combination of immuno-diffusion and electrophoresis. An antigen mixture is first separated into its components by electrophoresis and then tested by double immuno-diffusion.

Inflammation: A localized physical condition in which part of the body becomes reddened, swollen, hot, and often painful, especially as a reaction to injury or infection.

Innate immunity: It is the natural immunological, non-specific mechanism for fighting against infections.

Inoculation: In molecular biology, it is the process of introducing microbes into a culture media so that they reproduce there.

LB media: It is a rich medium containing tryptone, yeast extract and NaCl, that is commonly used to culture members of the Enterobacteriaceae as well as for coliphage plaque assays.

Ligation: It is the process of joining DNA fragments using a ligase enzyme.

Locus: A locus (plural loci) is a specific, fixed position on a chromosome where a particular gene or genetic marker is located.

Neutralization: The ability of antibodies to block the site(s) on bacteria or viruses that they use to enter their target cell.

Opsonisation: An immune response that uses opsonins (various extracellular proteins) to tag foreign pathogens for elimination by phagocytes.

Palindromic: It is a nucleic acid sequence in a double-stranded DNA or RNA molecule whereby reading in a certain direction on one strand is identical to the sequence in the same direction on the complementary strand.

Phagocytosis: The process by which a phagocyte (a type of WBC) surrounds and destroys foreign substances (such as bacteria) and removes dead cells.

Plasmid: It is a small, extrachromosomal DNA molecule within a cell that is physically separated from chromosomal DNA and can replicate independently.

Polymerase chain reaction: It (PCR) is a laboratory technique used to amplify DNA sequences.

Precipitin arc: Precipitin arcs form at the zone of equivalence between the antigen and specific antisera when they combine and precipitate.

Primers: It is a short single-stranded nucleic acid used by all living organisms in the initiation of DNA synthesis.

Probe: It is a single-stranded sequence of DNA or RNA used to search for its complementary sequence in a sample genome

Prozone: A zone containing high antibody concentration which interferes with the formation of antigen-antibody lattice causing a false negative response.

Radial Immunodiffusion: It is a quantitative method used in many clinical laboratories to determine the concentration of specific antigens or antibody classes (IgG, IgA, and IgM) in patient serum. It involves adding samples to circular holes cut in agar plates and the formation of a circular ring

Radio-Immunoassay: It is one of the sensitive immunoassay techniques which helps in the determination of antigens or antibodies in a sample with the use of radioisotopes.

Restriction endonuclease: It is a protein isolated from bacteria that cleaves DNA sequences after recognizing sequence-specific sites.

Restriction enzyme: Same as restriction endonuclease

Restriction fragment length polymorphism: It (abbreviated RFLP) refers to differences (or variations) among people in their DNA sequences at sites recognized by restriction enzymes.

Restriction map: A restriction map is a diagram that indicates the relative positions of restriction enzyme sites on a particular DNA sequence.

Restriction mapping: It is a physical mapping technique that is used to determine the relative location of restriction sites on a DNA fragment to give a restriction map.

Restriction-Modification system: It (RM system) is found in bacteria and other prokaryotic organisms, and provides a defense against foreign DNA, such as that borne by bacteriophages.

Rh factor: An inherited protein found on the surface of red blood cells. If someone's blood has the protein, he/she is Rh positive.

Rhogam: An injectable drug that is used to protect a Rh+ fetus from antibodies in an Rh- mother's blood and to prevent Rh allergy in the mother.

Satellite DNA: It (satDNA) is the highly repetitive DNA consisting of short sequences repeated a large number of times. It carries a variable AT-rich repeat unit that often forms arrays up to 100 Mb.

Serum: The clear liquid part of the blood that remains after blood cells and clotting proteins have been removed.

Colour Plates

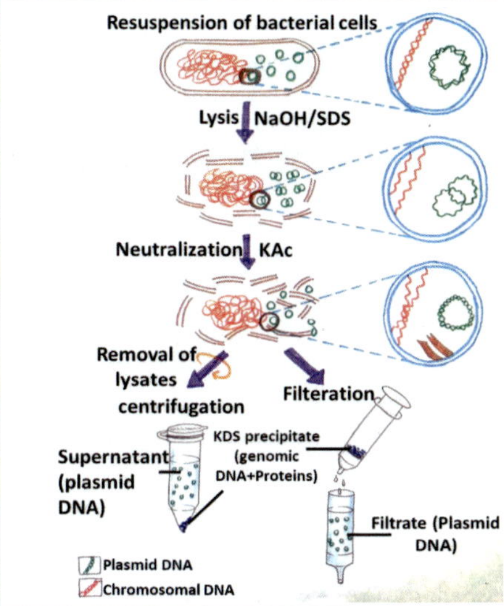

Figure 2a. Principle of Plasmid DNA Isolation (Illustrator: Aakriti Singh, Department of Botany, University of Delhi). (pg. 20)

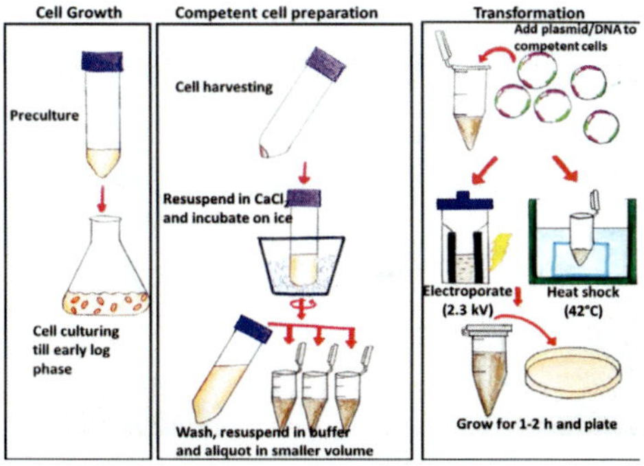

Figure 4a. Steps of Transformation Process (Illustrator: Aakriti Singh, Department of Botany, University of Delhi). (pg. 35)

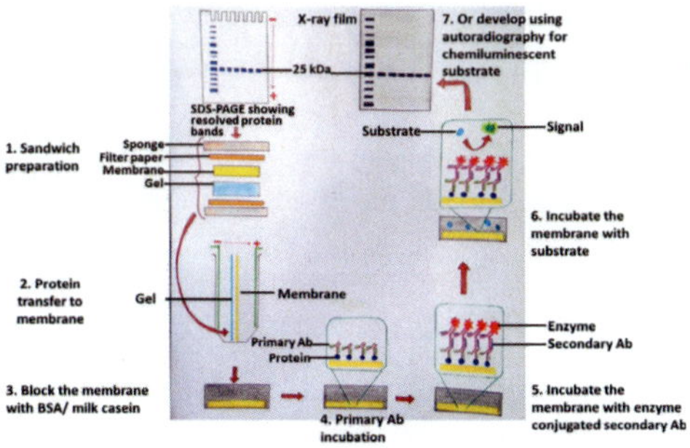

Figure 8a. Steps of Western Blotting Technique (Illustrator: Aakriti Singh, Department of Botany, University of Delhi). (pg. 72)

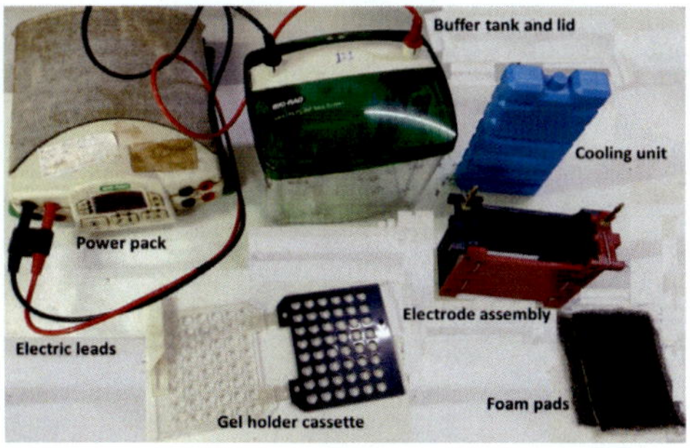

Figure 8b. Western Blot Equipment. (pg. 75)

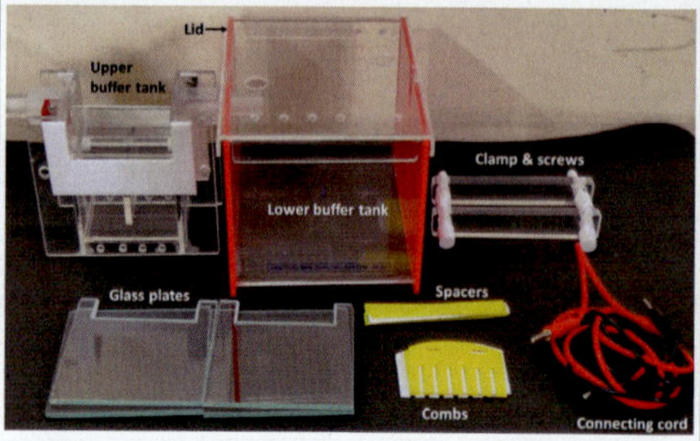

Figure 8c. SDS-PAGE Equipment Showing all the Parts (Pic courtesy: Dr. Nitin Bhardwaj). (pg. 75)

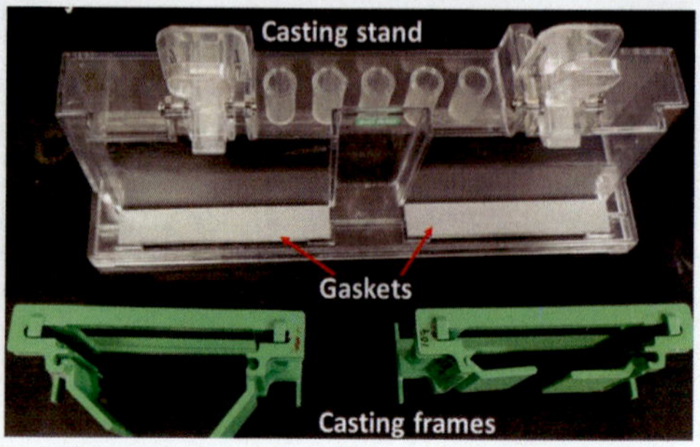

Figure 8d. Gel Casting Assembly. (pg. 76)

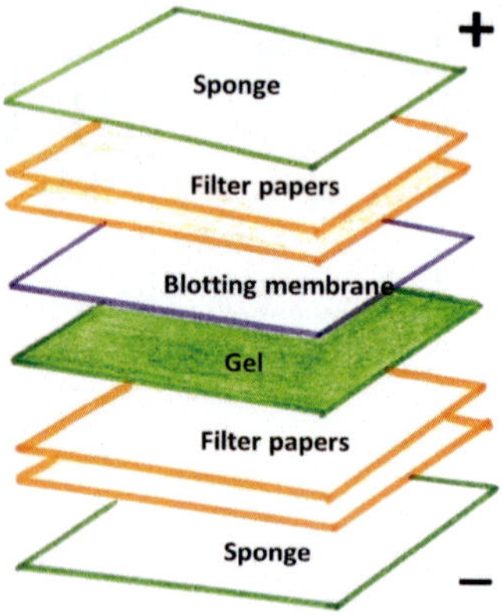

Figure 8e. The Sandwich Arrangement of Gel in Western Blotting (Illustrator: Aakriti Singh, Department of Botany, University of Delhi). (pg. 78)

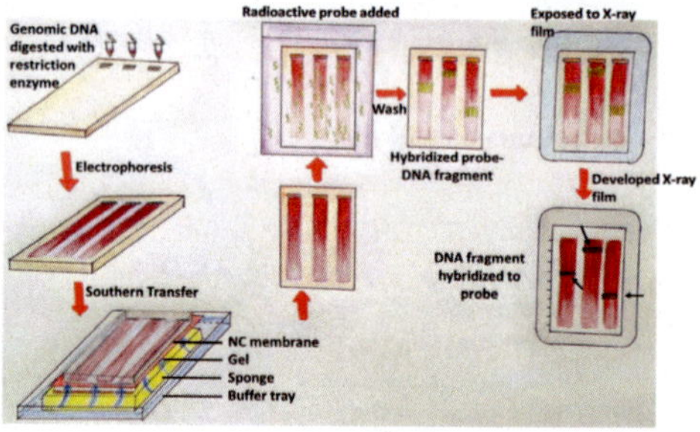

Figure 9a. Steps of Southern Blotting (Illustrator: Aakriti Singh, Department of Botany, University of Delhi) (Modified from Hood, 2006). (pg. 86)

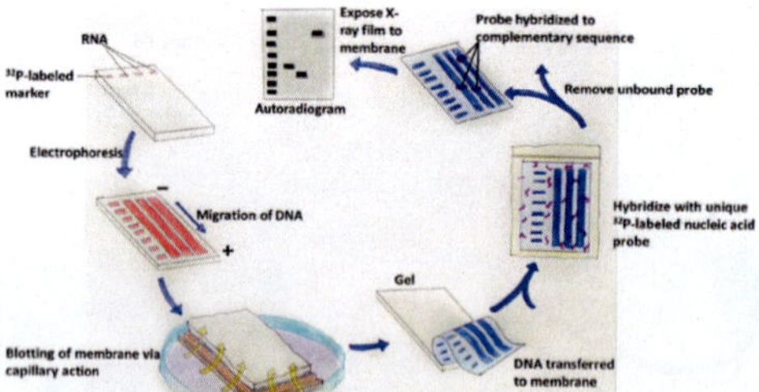

Figure 10a. Steps of Northern Blotting (Illustrator: Aakriti Singh, Department of Botany, University of Delhi). (pg. 93)

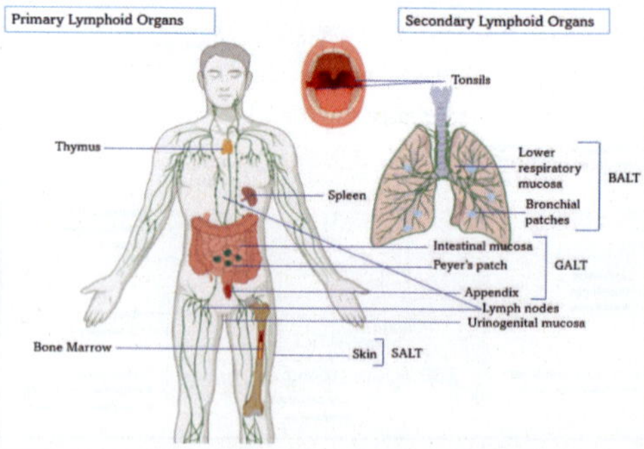

Figure 1a. Primary and Secondary Lymphoid Organs Present in Man. (pg. 101)

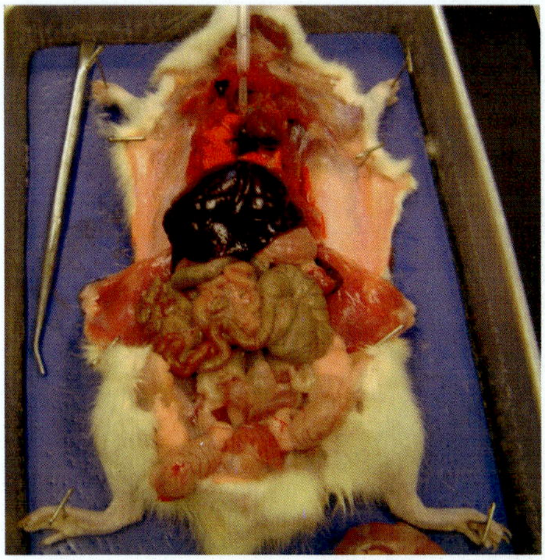

Figure 1b. Display of Anatomy of Dissected Rat. (*Source*: By Allen Lew from Berkeley, USA - Flickr, CC BY 2.0). (pg. 106)

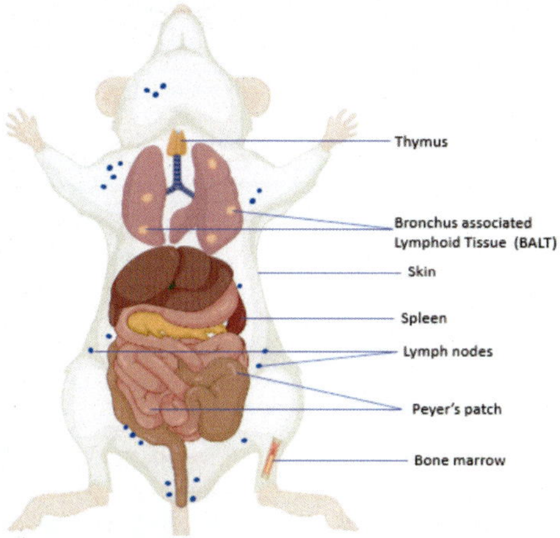

Figure 1c. Schematic Representation of Lymphoid Tissues in Rat. (pg. 107)

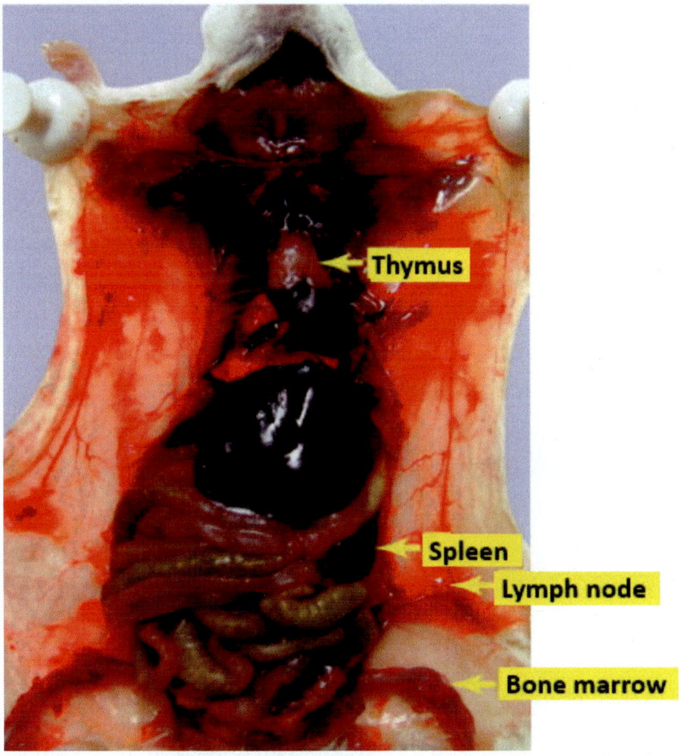

Figure 1d. Display of Lymphoid Organs in a Dissected Rat (*Source*: Mak ***et al.,*** 2014). (pg. 108)

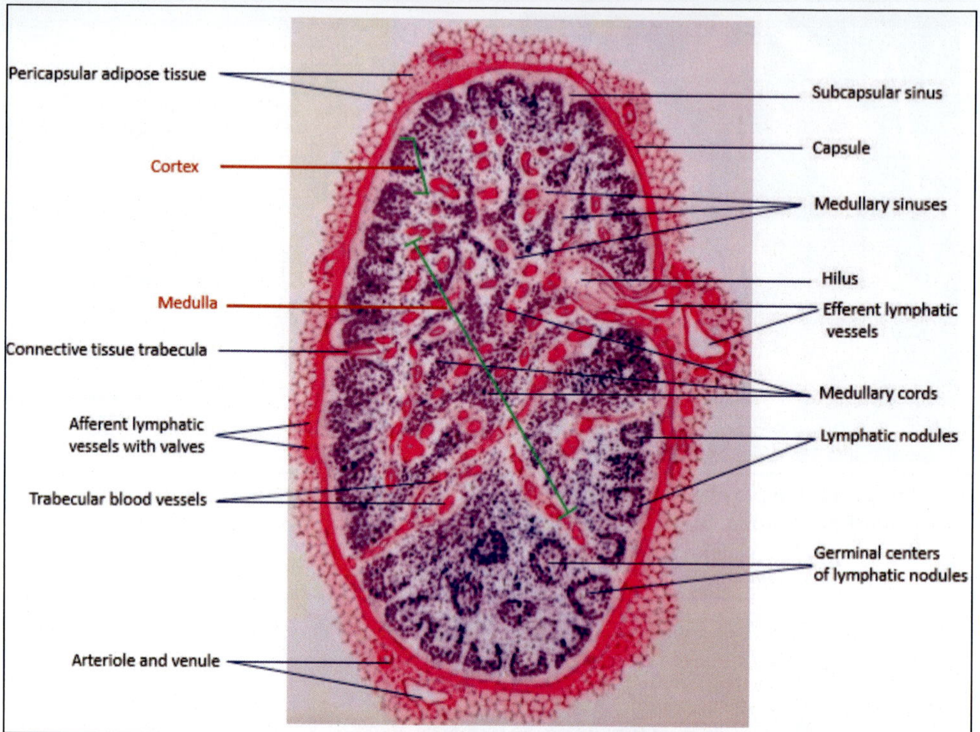

Figure 2a. T. S. of Lymph Node (Panoramic view): Stain: Hematoxylin and eosin (10×). (pg. 114)

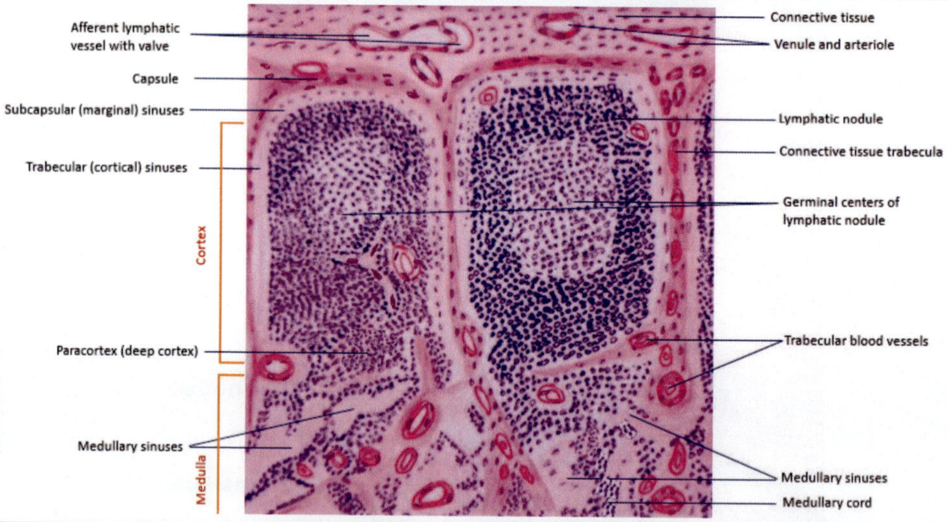

Figure 2b. T.S. of Lymph Node: Stain: hematoxylin and eosin. (40× view). (pg. 115)

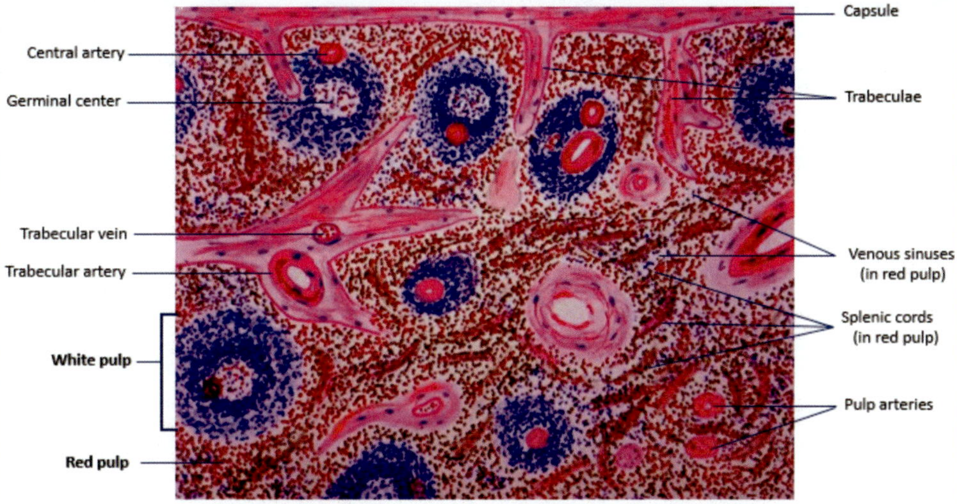

Figure 2c. T.S. of Spleen (Panoramic View). Stain: Hematoxylin and eosin. (10×). (pg. 117)

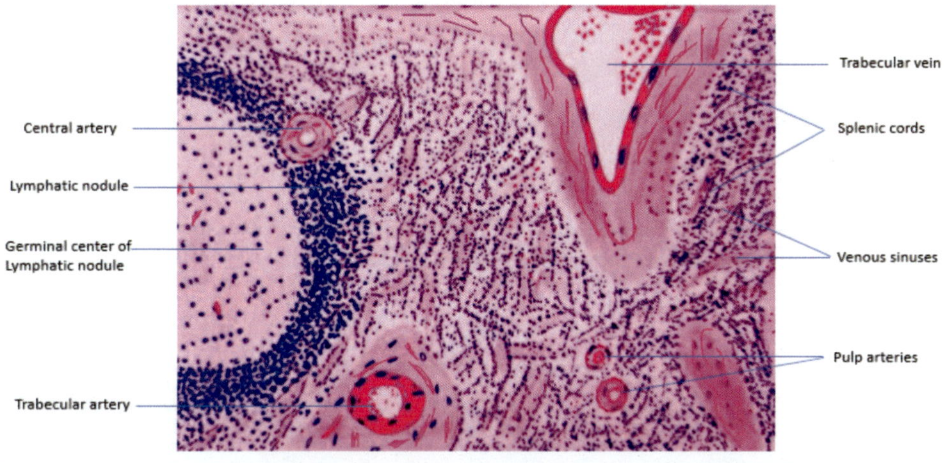

Figure 2d. T.S. of Spleen through Red and White Pulp. Stain: Hematoxylin and eosin. (40×). (pg. 118)

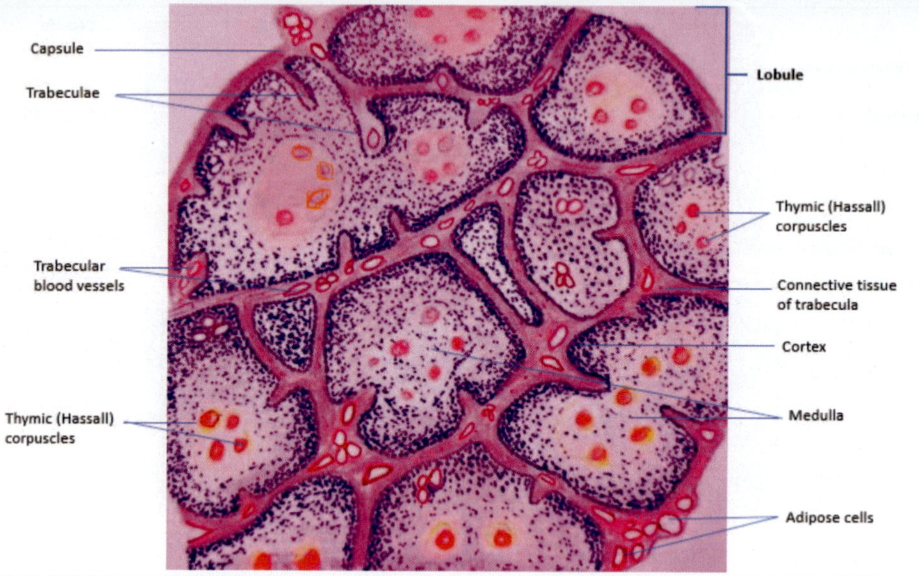

Figure 2e. T. S. of Thymus Gland (Panoramic View), Stain: Hematoxylin and eosin (10×). (pg. 120)

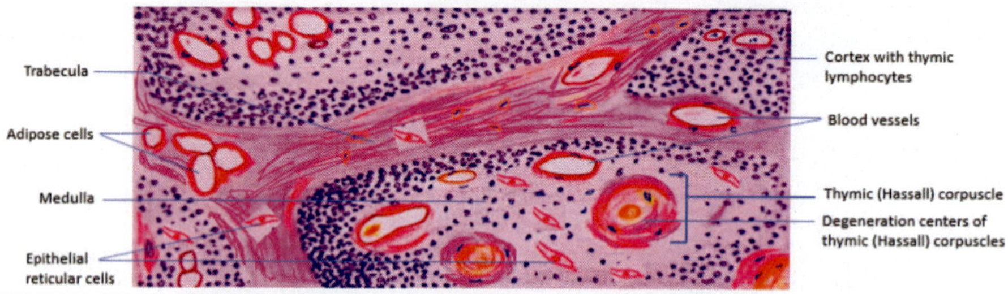

Figure 2f. T. S. of Thymus Gland. Stain: Hematoxylin and eosin (40×). (pg. 121)

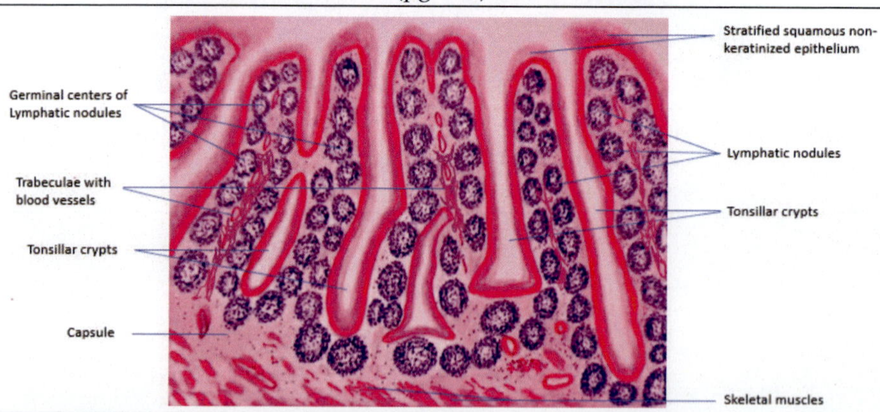

Figure 2g. T.S. of Palatine Tonsil. Stain: hematoxylin and eosin (10× view). (pg. 122)

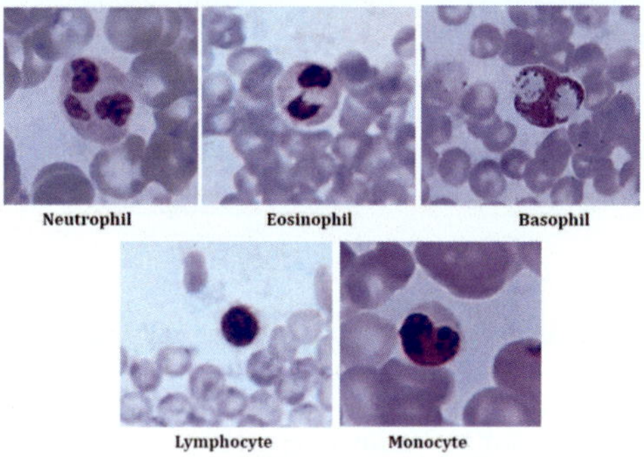

Figure 3b. Different Types of White Blood Cells (Stained with Giemsa stain)*. (pg. 130)

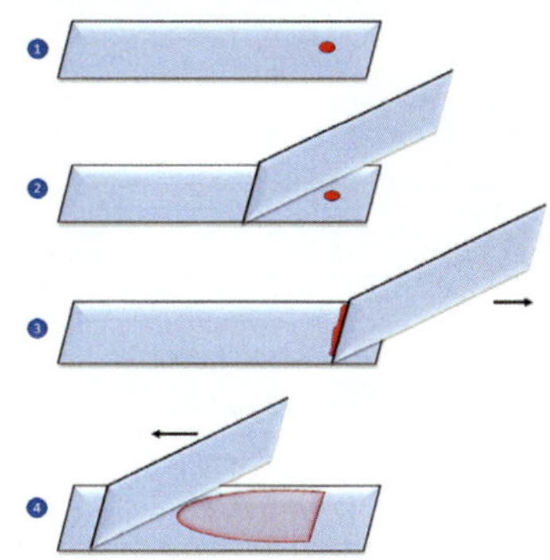

Figure 3c. Preparation of Thin Blood Smear. (pg. 133)

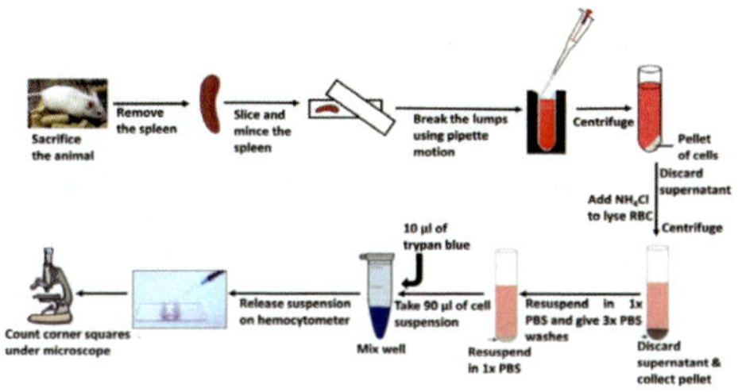

Figure 4a. Steps Showing Preparation of Splenocyte Suspension, Counting and Viability Test. (pg. 142)

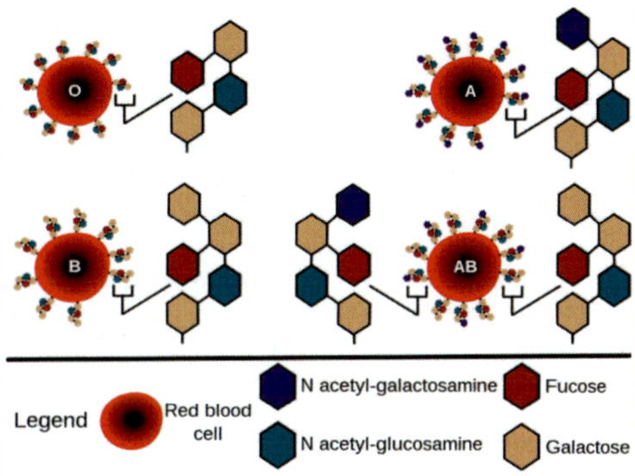

Figure 5a. Diagram Showing the Carbohydrate Chains that Determine the ABO Blood Group (***Source***: InvictaHOG, https://commons.wikimedia.org/w/index.php?curid=1161572). (pg. 148)

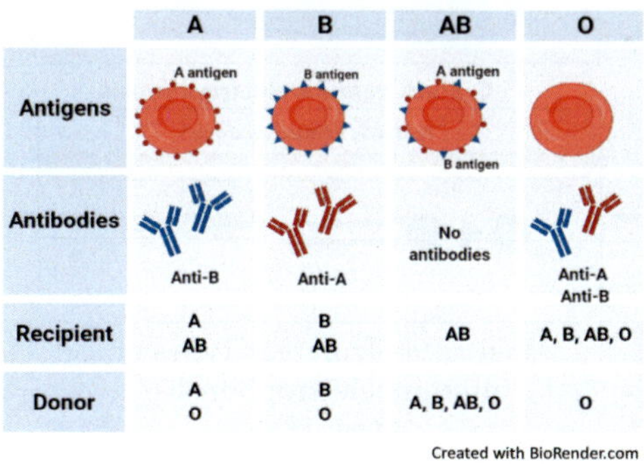

Figure 5b. Antigens and Antibodies Present in ABO Blood Group. (pg. 149)

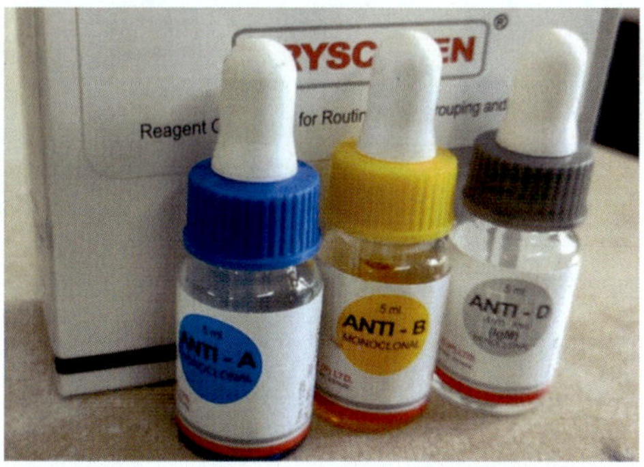

Figure 5d. A Kit Showing Anti-A, Anti-B and Anti-D Antibodies (Pic courtesy: Dr. Shailesh Kumar R. Yadav). (pg. 151)

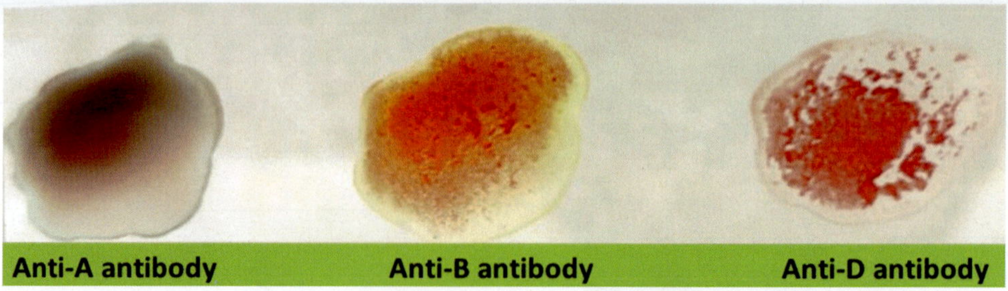

Figure 5e. Reaction of Antibodies to Blood Group. The slide shows B+ blood group reaction. (Pic courtesy: Dr. Shailesh Kumar R. Yadav). (pg. 151)

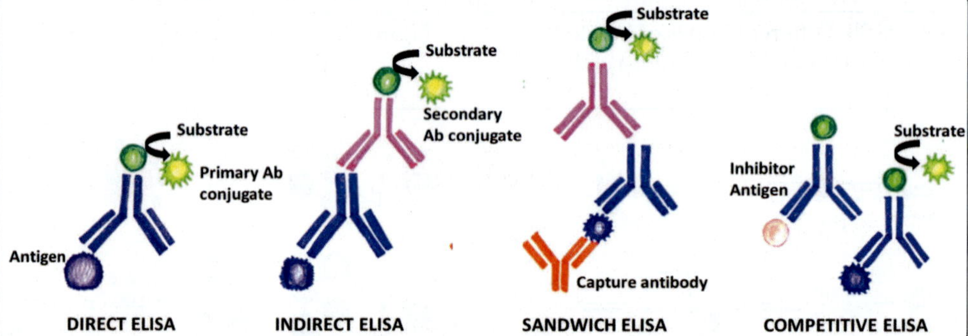

Figure 7b. Different Types of ELISA (Illustrator: Aakriti Singh, Department of Botany, University of Delhi). (pg. 166)

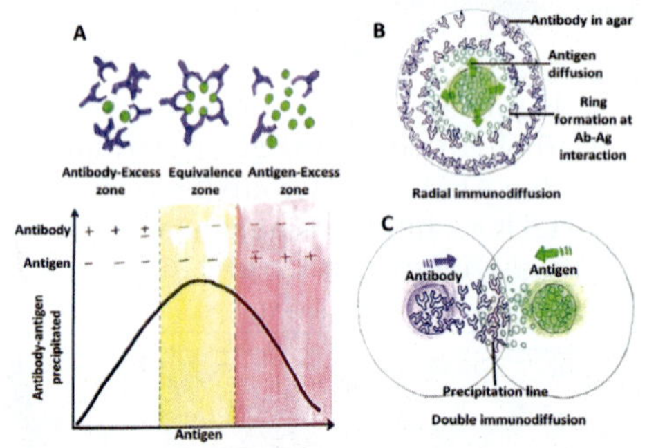

Figure 8b. (A) Precipitation curve showing various antibody and antigen zones. (B) Radial immunodiffusion (C) Double immunodiffusion (Illustrator: Aakriti Singh, Department of Botany, University of Delhi). (pg. 176)

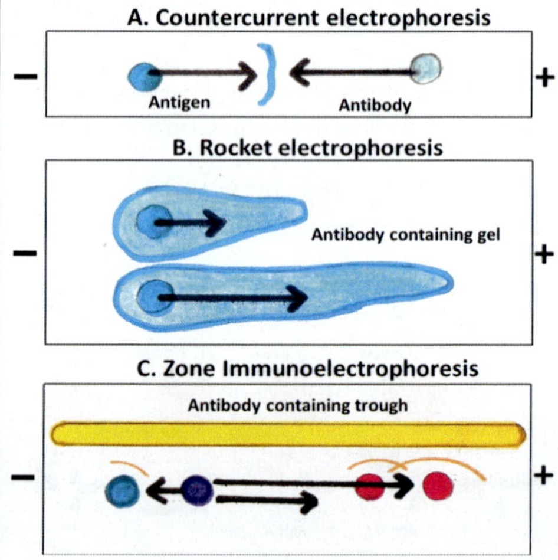

Figure 9b. Different Types of Immunoelectrophoresis. (pg. 183)

Splenocytes: It can be any one of the different white blood cell types as long as it is situated in the spleen or purified from splenic tissue

Supercoiled DNA: It refers to a higher-order DNA structure. The double-helical structure of DNA entails the interwinding of two complementary strands around one another and around a common helical axis.

***Taq* polymerase:** It is the heat-stable (thermostable) DNA polymerase extracted from the thermophilic bacteria *Thermus aquaticus* and most commonly used for PCR.

Template: It refers to the strand used by DNA polymerase or RNA polymerase to attach complementary bases during DNA replication or RNA transcription, respectively

Thawing: It is the process of warming the biological samples to control the rate of cooling and prevent the cell damage caused by the crystallization.

Thermal cycler: It (also known as a thermocycler, PCR machine or DNA amplifier) is a laboratory apparatus most commonly used to amplify segments of DNA.

Transduction: It is a mode of genetic transfer from one bacteria to another through a virus.

Transfection: It is a procedure that introduces foreign nucleic acids into cells to produce genetically modified cells.

Transformation efficiency: Transformation efficiency is defined as the number of colony forming units (CFU) per microgram of DNA used to transform the bacteria.

Transformation: It is one of several processes by which genetic material in the form of "naked" deoxyribonucleic acid (DNA) is transferred between microbial cells.

Universal Donor: A universal donor is a person who can donate blood to any recipient of any blood group. Blood group O- is considered a universal donor blood group.

Universal recipient: A universal recipient is a person who can receive blood from any donor of any blood group. Blood group AB+ is considered a universal recipient blood group.

Viable: It refers to the capability of cells to survive and live successfully under particular environmental conditions.

Index